Your *Clinics* subscription just got better!

You can now access the FULL TEXT of this publication online at no additional cost! Activate your online subscription today and receive...

- Full text of all issues from 2002 to the present
- Photographs, tables, illustrations, and references
- Comprehensive search capabilities
- Links to MEDLINE and Elsevier journals

Plus, you can also sign up for E-alerts of upcoming issues or articles that interest you, and take advantage of exclusive access to bonus features!

To activate your individual online subscription:

1. Visit our website at **www.TheClinics.com**.

2. Click on "Register" at the top of the page, and follow the instructions.

3. To activate your account, you will need your subscriber account number, which you can find on your mailing label (note: the number of digits in your subscriber account number varies from six to ten digits). See the sample below where the subscriber account number has been circled.

This is your subscriber account number

```
******************************************3-DIGIT 001
FEB00   J0167    C7    ( 123456-89 )  10/00   Q: 1

J.H. DOE, MD
531 MAIN ST
CENTER CITY, NY  10001-001
```

4. That's it! Your online a[...]urce for clinical reviews is now available.

theclinics.com

ELSEVIER

RADIOLOGIC CLINICS

of North America

Diagnostic Challenges in
Musculoskeletal Radiology

JAVIER BELTRAN, MD
Guest Editor

July 2005 • Volume 43 • Number 4

SAUNDERS

An Imprint of Elsevier, Inc.
PHILADELPHIA   LONDON   TORONTO   MONTREAL   SYDNEY   TOKYO

W.B. SAUNDERS COMPANY
A Division of Elsevier Inc.

1600 John F. Kennedy Boulevard • Suite 1800 • Philadelphia, Pennsylvania 19103-2899

http://www.theclinics.com

RADIOLOGIC CLINICS OF NORTH AMERICA
July 2005
Editor: Barton Dudlick

Volume 43, Number 4
ISSN 0033-8389
ISBN 1-4160-2761-0

Reprints: For copies of 100 or more, of articles in this publication, please contact the Commercial Reprints Department, Elsevier Inc., 360 Park Avenue South, New York, New York 10010-1710. Tel.: (+1) 212-633-3813; Fax: (+1) 212-462-1935; E-mail: reprints@elsevier.com.

The ideas and opinions expressed in *Radiologic Clinics of North America* do not necessarily reflect those of the Publisher. The Publisher does not assume any responsibility for any injury and/or damage to persons or property arising out of or related to any use of the material contained in this periodical. The reader is advised to check the appropriate medical literature and the product information currently provided by the manufacturer of each drug to be administered to verify the dosage, the method and duration of administration, or contraindications. It is the responsibility of the treating physician or other health care professional, relying on independent experience and knowledge of the patient, to determine drug dosages and the best treatment for the patient. Mention of any product in this issue should not be construed as endorsement by the contributors, editors, or the Publisher of the product or manufacturers' claims.

Radiologic Clinics of North America (ISSN 0033-8389) is published bimonthly by W.B. Saunders Company. Corporate and editorial offices: 1600 John F. Kennedy Boulevard, Suite 1800, Philadelphia, Pennsylvania 19103-2899. Accounting and circulation offices: 6277 Sea Harbor Drive, Orlando, FL 32887-4800. Periodicals postage paid at Orlando, FL 32862, and additional mailing offices. Subscription prices are USD 220 per year for US individuals, USD 331 per year for US institutions, USD 110 per year for US students and residents, USD 255 per year for Canadian individuals, USD 405 per year for Canadian institutions, USD 299 per year for international individuals, USD 405 per year for international institutions and USD 150 per year for Canadian and foreign students/residents. To receive student and resident rate, orders must be accompanied by name of affiliated institution, date of term, and the *signature* of program/residency coordinator on institution letterhead. Orders will be billed at individual rate until proof of status is received. Foreign air speed delivery is included in all *Clinics* subscription prices. All prices are subject to change without notice. POSTMASTER: Send address changes to *Radiologic Clinics of North America*, W.B. Saunders Company, Periodicals Fulfillment, Orlando, FL 32887-4800. **Customer Service: 800-654-2452 (US). From outside of the US, call (+1) 407-345-4000.**

Radiologic Clinics of North America also is published in Greek by Paschalidis Medical Publications, Athens, Greece.

Radiologic Clinics of North America is covered in *Index Medicus, EMBASE/Excerpta Medica, Current Contents/Life Sciences, Current Contents/Clinical Medicine, RSNA Index to Imaging Literature, BIOSIS, Science Citation Index, and ISI/BIOMED.*

Printed in the United States of America.

GOAL STATEMENT

The goal of the *Radiologic Clinics of North America* is to keep practicing radiologists and radiology residents up to date with current clinical practice in radiology by providing timely articles reviewing the state of the art in patient care.

ACCREDITATION

The *Radiologic Clinics of North America* is planned and implemented in accordance with the Essential Areas and Policies of the Accreditation Council for Continuing Medical Education (ACCME) through the joint sponsorship of the University of Virginia School of Medicine and Elsevier. The University of Virginia School of Medicine is accredited by the ACCME to provide continuing medical education for physicians.

The University of Virginia School of Medicine designates this educational activity for a maximum of 90 category 1 credits per year, 15 category 1 credits per issue, toward the AMA Physician's Recognition Award. Each physician should claim only those credits that he/she actually spent in the activity.

The American Medical Association has determined that physicians not licensed in the US who participate in this CME activity are eligible for AMA PRA category 1 credit.

AMA PRA category 1 credit can be earned by reading the text material, taking the examination online at http://www.theclinics.com/home/cme, and completing the evaluation. After taking the test, your will be required to review any and all incorrect answers. Following completion of the test and the evaluation, your credit will be awarded and you may print your certificate.

FACULTY DISCLOSURE

As a provider accredited by the Accreditation Council for Continuing Medical Education (ACCME), the Office of Continuing Medical Education of the University of Virginia School of Medicine must ensure balance, independence, objectivity, and scientific rigor in all its individually sponsored or jointly sponsored educational activities. All authors/editors participating in a sponsored activity are expected to disclose to the readers any significant financial interest or other relationship (1) with the manufacturer(s) of any commercial product(s) and/or provider(s) of commercial services discussed in an educational presentation and (2) with any commercial supporters of the activity (significant financial interest or other relationship can include such things as grants or research support, employee, consultant, stock holder, member of speakers bureau, etc.) The intent of this disclosure is not to prevent authors/editors with a significant financial or other relationship from writing an article, but rather to provide readers with information on which they can make their own judgments. It remains for the readers to determine whether the author's/editor's interest or relationships may influence the article with regard to exposition or conclusion.

The authors/editors listed below have identified no professional or financial affiliations related to their article:
Faustino Abascal, MD; Ronald S. Adler, MD, PhD; Martiza Angulo, MD; Javier Beltran, MD, FACR; Ron Boucher, MD; Ana Canga, MD; Luis Cerezal, MD; Deep S. Chatha, MD; Qi Chen, MD; Christine B. Chung, MD; Anne Cotten, MD; Patricia M. Cunningham, MD; Francisco del Piñal, MD; Barton Dudlick, Acquisitions Editor; Christine Galant, MD; Roberto García-Valtuille, MD; Christian Glaser, MD; Tudor Hughes, MD; Ilma L. Isaza, MD; Marlena Jbara, MD; Philipp Lang, MD, MBA; Frédéric E. Lecouvet, MD, PhD; Baudouin E. Maldague, MD; Jacques Malghem, MD; Paul Marten, MD; Morcos Morcos, MD; Farimah Noorbakhsh, MD; Mark E. Schweitzer, MD; Thierry Tavernier, MD; Bruno C. Vande Berg, MD, PhD; and, Hiroshi Yoshioka, MD, MSc.

Disclosure of discussion of non-FDA approved uses for pharmaceutical products and/or medical devices:
The University of Virginia School of Medicine, as an ACCME provider, requires that all authors/editors identify and disclose any "off label" uses for pharmaceutical products and/or for medical devices. The University of Virginia School of Medicine recommends that each reader fully review all the available data on new products or procedures prior to instituting them with patients.

All authors/editors who provided disclosures will not be discussing any off-label uses except:
Christine B. Chung, MD will discuss the use of gadolinium for MR arthrography.

The following authors have not provided disclosure or off-label information.
Kathleen C. Finzel, MD.

TO ENROLL

To enroll in the Radiologic Clinics of North America Continuing Medical Education program, call customer service at 1-800-654-2452 or sign up online at http://www.theclinics.com/home/cme. The CME program is available to subscribers for an additional annual fee of USD 195.

FORTHCOMING ISSUES

RECENT ISSUES

GUEST EDITOR

JAVIER BELTRAN, MD, FACR, Chairman, Department of Radiology, Maimonides Medical Center; Clinical Professor, Radiology, State University of New York Downstate, Brooklyn; and Professor, Clinical Radiology, Mount Sinai School of Medicine, New York, New York

CONTRIBUTORS

FAUSTINO ABASCAL, MD, Radiologist, Department of Radiology, Instituto Radiológico Cántabro, Clínica Mompía, Mompía, Cantabria, Spain

RONALD S. ADLER, MD, PhD, Attending Radiologist and Chief, Division of Ultrasound and Body Imaging, Department of Radiology and Imaging, Hospital for Special Surgery; and Professor, Weill Medical College of Cornell University, New York, New York

MARITZA ANGULO, MD, Staff Radiologist, Fundacion Clinica Medica Sur, Delegacion Tlalpan, Mexico

JAVIER BELTRAN, MD, FACR, Chairman, Department of Radiology, Maimonides Medical Center; Clinical Professor, Radiology, State University of New York Downstate, Brooklyn; and Professor, Clinical Radiology, Mount Sinai School of Medicine, New York, New York

RON BOUCHER, MD, Assistant Chairman, Radiology; and Academic Chief, MR Imaging, Naval Medical Center, San Diego, California

ANA CANGA, MD, Consultant Radiologist, Department of Radiology, Instituto Radiológico Cántabro, Clínica Mompía, Mompía, Cantabria, Spain

LUIS CEREZAL, MD, Head, Department of Radiology, Instituto Radiológico Cántabro, Clínica Mompía, Mompía, Cantabria, Spain

DEEP S. CHATHA, MD, Department of Radiology, Hospital for Joint Diseases Orthopaedic Institute, New York, New York

QI CHEN, MD, Resident, Department of Radiology, Maimonides Medical Center, Brooklyn, New York

CHRISTINE B. CHUNG, MD, Assistant Professor, Department of Radiology, University of California at San Diego and Veterans Affairs Healthcare System, La Jolla, California

ANNE COTTEN, MD, Professor and Head, Service de Radiologie Ostéo-Articulaire, Hôpital Roger Salengro, Lille, France

PATRICIA M. CUNNINGHAM, MD, Department of Radiology, Hospital for Joint Diseases Orthopaedic Institute, New York, New York

FRANCISCO DEL PIÑAL, MD, Head, Department of Private Hand–Wrist and Plastic–Reconstructive Surgery and Hand Surgery Department, Mutua Montañesa, Santander, Spain

KATHLEEN C. FINZEL, MD, Associate Attending Radiologist, Department of Radiology and Imaging; and Assistant Professor, Radiology, Weill Medical College of Cornell University, New York, New York

CHRISTINE GALANT, MD, Associate Professor, Saint Luc University Hospital, Université de Louvain, Brussels, Belgium

ROBERTO GARCÍA-VALTUILLE, MD, Radiologist, Department of Radiology, Instituto Radiológico Cántabro, Clínica Mompía, Mompía, Cantabria, Spain

CHRISTIAN GLASER, MD, Consultant, Musculoskeletal Imaging; and Section Chief, Division of General Radiography, Department of Clinical Radiology, Ludwig-Maximilians-Universität München, Munich, Germany

TUDOR HUGHES, MD, Associate Professor, Department of Radiology, University of California at San Diego and Veterans Affairs Healthcare System, La Jolla, California

ILMA L. ISAZA, MD, Staff Radiologist, CT Scanner de Mexico, Colonia Roma, Mexico

MARLENA JBARA, MD, Assistant Clinical Professor, Maimonides Medical Center, Brooklyn, New York

PHILIPP LANG, MD, MBA, Associate Professor and Director, Division of Musculoskeletal Radiology, Department of Radiology, Brigham and Women's Hospital, Harvard Medical School, Boston, Massachusetts

FRÉDÉRIC E. LECOUVET, MD, PhD, Professor, Section of Musculoskeletal Radiology, Department of Radiology, Saint Luc University Hospital, Université de Louvain, Brussels, Belgium

BAUDOUIN E. MALDAGUE, MD, Professor, Section of Musculoskeletal Radiology, Department of Radiology, Saint Luc University Hospital, Université de Louvain, Brussels, Belgium

JACQUES MALGHEM, MD, Professor, Section of Musculoskeletal Radiology, Department of Radiology, Saint Luc University Hospital, Université de Louvain, Brussels, Belgium

PAUL MARTEN, MD, Resident, Department of Radiology, Maimonides Medical Center, Brooklyn, New York

MORCOS MORCOS, MD, Resident, Department of Radiology, Maimonides Medical Center, Brooklyn, New York

FARIMAH NOORBAKHSH, MD, Visiting Research Fellow, Division of Musculoskeletal Radiology, Department of Radiology, Brigham and Women's Hospital, Harvard Medical School, Boston, Massachusetts

MARK E. SCHWEITZER, MD, Department of Radiology, Hospital for Joint Diseases Orthopaedic Institute, New York, New York

THIERRY TAVERNIER, MD, Radiologist, Imagerie Médicale, Clinique de la Sauvegarde, Lyon, France

BRUNO C. VANDE BERG, MD, PhD, Professor and Chief, Section of Musculoskeletal Radiology, Department of Radiology, Saint Luc University Hospital, Université de Louvain, Brussels, Belgium

HIROSHI YOSHIOKA, MD, MSc, Assistant Professor, Division of Musculoskeletal Radiology, Department of Radiology, Brigham and Women's Hospital, Harvard Medical School, Boston, Massachusetts

CONTENTS

Osteoarthritis is the most common type of arthritis and a frequent cause of pain and disability. A number of exciting surgical treatment modalities have been introduced recently, including autologous chondrocyte transplantation and osteochondral allografting or autografting. MR imaging offers the distinct advantage of visualizing the articular cartilage directly. MR imaging can detect signal and morphologic changes in the cartilage and has been used to detect cartilage surface fraying, fissuring, and varying degrees of cartilage thinning.

In view of recent therapeutic approaches to cartilage damage in osteoarthritis, it is necessary to develop and further refine noninvasive quantitative tools for specific diagnosis and follow-up studies. There is considerable experimental and some clinical experience with T2 relaxation time measurements. Motivation for diffusion-weighted imaging and diffusion-tensor imaging as comparably new techniques for cartilage imaging is to obtain directly additional three-dimensional architectural and directional information about the cartilage matrix.

This article focuses on spontaneous painful conditions involving the subchondral bone and marrow of mature knee epiphyses. MR imaging is the technique of choice for the work-up of these lesions and enables distinction of two main categories of lesions on the basis of T1-weighted images: avascular necrosis and lesions presenting the bone marrow edema pattern. This latter category encompasses spontaneous osteonecrosis of the knee, and a variety of self-resolving conditions that may be differentiated by the study of the subchondral bone marrow area on T2-weighted images. Behind definite appellation of lesions, the challenge for the radiologist is to provide a prognosis: the distinction between self-resolving lesions from those that may evolve to epiphyseal collapse and joint impairment should be possible in most cases.

High- Versus Low-Field MR Imaging 673
Thierry Tavernier and Anne Cotten

The role of MR imaging as a noninvasive technique in the detection and evaluation of musculoskeletal diseases is unquestionable. Most of the studies reported in the literature are based on high-field MR imaging. Initial studies performed with low-field-strength have reported unsatisfactory results in the assessment of the musculoskeletal system. Recent improvements, however, have generated a renewed interest in low-field-strength MR imaging. This article presents the principal applications and results published in the literature.

Shoulder MR Arthrography: How, Why, When 683
Marlena Jbara, Qi Chen, Paul Marten, Morcos Morcos, and Javier Beltran

This article reviews current MR techniques for shoulder imaging, discusses advantages and disadvantages of each, and reviews the literature regarding sensitivity, specificity, and accuracy of MR arthrography versus nonenhanced MR in the evaluation of shoulder pathology, specifically, glenoid labral and rotator cuff tears.

Ankle MR Arthrography: How, Why, When 693
Luis Cerezal, Faustino Abascal, Roberto García-Valtuille, and Ana Canga

MR arthrography has become an important tool for the assessment of a variety of ankle disorders. MR arthrography may facilitate the evaluation of patients with suspected intra-articular pathology in whom conventional MR imaging is not sufficient for an adequate diagnosis and be useful for therapy planning. MR arthrography is valuable in the evaluation of ligamentous injuries, impingement syndromes, cartilage lesions, osteochondral lesions of the talus, loose bodies, and several synovial joint disorders. Indirect MR arthrography is a useful adjunct to conventional MR imaging and may be preferable to direct MR arthrography in cases in which an invasive procedure is contraindicated or when fluoroscopy is not available.

Wrist MR Arthrography: How, Why, When 709
Luis Cerezal, Faustino Abascal, Roberto García-Valtuille, and Francisco del Piñal

MR imaging of the wrist frequently represents a diagnostic challenge for radiologists because of the complex anatomy of the wrist joint, the small size of its components, and little known pathologic conditions. MR arthrography combines the advantages of conventional MR imaging and arthrography by improving the visualization of small intra-articular abnormalities. This article reviews the current role of MR arthrography in the evaluation of wrist joint disorders considering the relevant aspects of anatomy, techniques, and applications.

MR Arthrography of the Knee: How, Why, When 733
Christine B. Chung, Ilma L. Isaza, Maritza Angulo, Ron Boucher, and Tudor Hughes

MR arthrography combines the techniques of arthrography with MR imaging to benefit from the added imaging information afforded by intra-articular distention. This article reviews technical considerations for MR arthrography, potential complications, indications, pitfalls in imaging diagnosis, and commonly encountered pathology. It is an elegant study that can offer precise diagnostic information in the appropriate clinical setting.

ELSEVIER
SAUNDERS

Radiol Clin N Am 43 (2005) xi

RADIOLOGIC
CLINICS
of North America

Preface

Diagnostic Challenges in Musculoskeletal Radiology

Javier Beltran, MD
Guest Editor

Conventional radiology has been used from its inception in the late nineteenth century for the diagnosis of bone lesions, and its strengths and weaknesses have been recognized for a long time. CT, which was developed in the mid-1970s, was used extensively for the evaluation of musculoskeletal conditions until MR imaging almost entirely replaced it in the late 1980s. Ultrasound has been slowly gaining strength as a useful tool in musculoskeletal imaging, mostly in Europe, Australasia, Canada, and South America, and it is now finally becoming more popular in the United States. With the advancement of CT technology and the development of multidetector technology, we are witnessing a comeback of CT for the evaluation of musculoskeletal conditions. Although many entities are now well evaluated with these technologies, some of them still represent a challenge to the interpreting radiologist.

In this issue of the *Radiologic Clinics of North America*, the authors discuss areas of the musculoskeletal system in which technical or interpretative challenges still exist. Given the widespread use of MR imaging, most of the articles deal with MR imaging issues, including MR arthrography of different joints, cartilage imaging, and controversies related to field strength. An additional article outlines the usage of ultrasound for the assessment of common and less common lesions involving the tendons, and two articles discuss specific disease entities, including the diabetic foot and epiphyseal lesions of the knee.

I want to thank the authors for their time and effort in contributing to a very up-to-date review of current concepts and in outlining the imaging and interpretative controversies.

Javier Beltran, MD
Chairman
Department of Radiology
Maimonides Medical Center
4802 Tenth Avenue
Brooklyn, NY 11219, USA
E-mail address: javierb@radreports.com

RADIOLOGIC CLINICS
of North America

Radiol Clin N Am 43 (2005) 629 – 639

MR Imaging of Articular Cartilage: Current State and Recent Developments

Philipp Lang, MD, MBA*, Farimah Noorbakhsh, MD, Hiroshi Yoshioka, MD, MSc

Division of Musculoskeletal Radiology, Department of Radiology, Brigham and Women's Hospital, Harvard Medical School, 75 Francis Street, Boston, MA 02115, USA

Osteoarthritis (OA) is the most common type of arthritis and a frequent cause of pain and disability [1]. A number of exciting surgical treatment modalities have been introduced recently, including autologous chondrocyte transplantation [2,3] and osteochondral allografting [4,5] or autografting [6].

Conventional radiography is widely used in evaluating the long-term progression of OA and is able clearly to depict the established hallmarks of OA, namely joint space narrowing, subchondral sclerosis, subchondral cyst formation, and osteophytosis [7,8]. Conventional radiography is limited, however, by its inability directly to visualize articular cartilage, the tissue in which the earliest insults of OA are thought to occur [9]. Radiographic measurements of joint space width cannot differentiate between femoral and tibial cartilage loss and do not reveal the distribution pattern of tissue degradation throughout the joint surface [9]. Moreover, highly standardized positioning procedures and even fluoroscopic control of the exact position of the joint are required to obtain reproducible data on joint space narrowing, which is used as a surrogate measure of cartilage degeneration and disease progression [7,8].

MR imaging offers the distinct advantage of visualizing the articular cartilage directly. MR imaging can detect signal and morphologic changes in the cartilage and has been used to detect cartilage surface fraying, fissuring, and varying degrees of cartilage thinning [10–16].

Standard MR imaging pulse sequences

The standard techniques broadly used in clinical practice and scientific studies are the two-dimensional fast spin echo (FSE) (Fig. 1) and the three-dimensional spoiled gradient-echo (SPGR) sequence [13,17]. Both sequences are available on most MR imaging systems.

Two-dimensional fast spin-echo imaging

FSE imaging affords high contrast for evaluating articular disorders and cartilage (Fig. 2) [10–13, 17,18]. Incidental magnetization transfer contrast contributes to the signal characteristics of articular cartilage on FSE images and can enhance the contrast between cartilage and joint fluid. Two-dimensional FSE sequences have excellent signal-to-noise ratios, which help to achieve short scan times in clinical practice. The sequence has fewer artifacts than three-dimensional SPGR [19]. Image blurring can be a problem in two-dimensional FSE. Strategies to decrease or avoid image blurring include the use of ultrashort echo times and short echo trains [20,21].

Three-dimensional spoiled gradient echo imaging

SPGR sequences have been used because of their ability to provide high-resolution three-dimensional images [10–12,18]. Fat suppression is typically used

This article is supported by National Institutes of Health/National Institute of Arthritis and Musculoskeletal and Skin Diseases Grant 1 RO1 AR051873-01.

* Corresponding author.
E-mail address: pklang@partners.org (P. Lang).

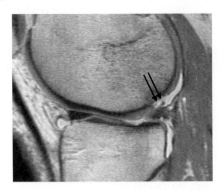

Fig. 1. Standard two-dimensional FSE pulse sequence. A sagittal MR image demonstrates the focal cartilage defect at the posterior aspect of the femoral condyle (*arrows*).

to increase the dynamic range of signal intensities in cartilage. The hyaline cartilage appears as a high signal intensity structure compared with adjacent tissues, which demonstrate lower signal intensities with this sequence. The three-dimensional imaging capability of this sequence has helped transform it into the standard acquisition technique for quantitative cartilage assessment, such as three-dimensional volume or thickness measurements. Recent studies indicate, however, that this sequence is hampered by significant image artifacts that can result in overestimation or underestimation of cartilage disease and failure of automated cartilage segmentation for three-dimensional analysis because of poor contrast between cartilage and surrounding tissues [22].

Many other MR imaging sequences have been proposed for cartilage imaging, but have not found widespread acceptance. These include T1-weighted [23–26], proton density weighted, and T2-weighted spin echo sequences [14,27,28]; inversion recovery sequences [29]; two- and three-dimensional magnetization transfer contrast sequences [16,30,31]; projection reconstruction spectroscopic imaging [32–34]; and two- and three-dimensional driven equilibrium Fourier transform (DEFT) [35–39]. Poor cartilage signal-to-noise (SNR) and contrast-to-noise ratios (spin echo, inversion recovery sequences), limited SNR efficiency (spin echo, inversion recovery), and need for off-line reconstruction (projection reconstruction spectroscopic imaging) or for image subtraction (magnetization transfer contrast) are among the factors that have prevented the broad dissemination and acceptance of these techniques for cartilage MR imaging.

The most promising novel MR imaging pulse sequences for cartilage imaging are water-selective excitation techniques, such as three-dimensional

SPGR with spectral spatial pulses (SS-SPGR) (ie, a water-selective excitation gradient echo sequence) [40]; three-dimensional steady-state free precession (SSFP) [35,41]; and three-dimensional FSE techniques [20,21,42]. These fast sequences hold the promise of providing three-dimensional coverage (unlike two-dimensional FSE) while yielding superior contrast-to-noise ratio between cartilage and surrounding tissues (unlike three-dimensional SPGR) and are likely to improve the accuracy and reproducibility of cartilage MR imaging.

Sensitivity and specificity of MR imaging

The sensitivity and specificity of standard MR in imaging detecting cartilage loss has been examined by correlating two-dimensional FSE or three-dimensional SPGR sequences with arthroscopic findings [10,11,13,14,17,18,43,44]. The specificity of standard two-dimensional FSE and three-dimensional SPGR sequences is excellent, ranging between 81% and 97% [10,11,14,17,18,43,44]. The data reported on the sensitivity of two-dimensional FSE [13,43] and three-dimensional SPGR [10–12,14] sequences in detecting cartilage loss are inconsistent, ranging between 60% and 94% [10,11,14,17,18,43,44].

With three-dimensional SPGR, image artifacts and a poorly defined cartilage surface contour along the posterior femoral condyle can result in false-negative or false-positive results, which may account for some of the reported variability in sensitivity. The severity of cartilage loss and the grade of OA are also important. Kawahara et al [44] reported that the

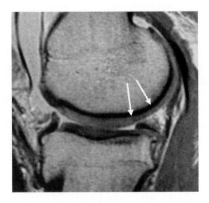

Fig. 2. Standard two-dimensional FSE pulse sequences. A sagittal MR image demonstrates diffuse, slightly less than 50% of cartilage thickness thinning along the central to posterior aspect of the medial femoral condyle (*arrows*). This finding is consistent with early cartilage loss in osteoarthritis.

sensitivity of two-dimensional FSE improved with higher grades of cartilage loss; the sensitivity reported in this study for superficial cartilage lesions was only 31.8%, whereas the sensitivity for full-thickness defects was greater than 90% [44]. Limited spatial resolution of the two-dimensional FSE sequence in slice direction may be the cause for this observation. Bredella et al [43] reported a sensitivity of only 61% for single-plane fat-saturated two-dimensional FSE sequences; when two or more planes were combined in the interpretation (eg, axial and coronal plane), the sensitivity increased to 93%. These data along with the limited sensitivity observed for superficial cartilage lesions in the study by Kawahara et al [44] provide a strong indication that pulse sequences with near isotropic resolution, such as three-dimensional SSFP or three-dimensional FSE, are needed to achieve sensitivities of cartilage MR imaging that are consistently greater than 90%.

Reproducibility of standard cartilage MR imaging pulse sequences

The reported reproducibility of visual readings of cartilage MR imaging acquired with a standard three-dimensional SPGR sequence is fair. In one study involving independent readings of 30 OA patients by three radiologists, the median interobserver agreement was 0.29 (range 0.06–0.38) [45]. The authors concluded that this was likely the result of errors related to partial volume averaging near the intercondylar notch and other image artifacts. This observation emphasizes the need for novel, more reproducible cartilage-sensitive imaging sequences.

Three-dimensional measurements of total cartilage volume and cartilage thickness have evolved as the standard for quantitative MR imaging–based assessment of cartilage loss [15,22,46–60]. Both measurements require segmentation of the cartilage from the surrounding tissue using such techniques as manual segmentation [15,22,52], signal intensity–based thresholding [15,52], seed-growing algorithms [53], filtering [61–63], watershed [64] and live wire approaches [54,55,65,66], or model-based segmentation [57,67–69]. Three-dimensional thickness maps can be generated using a three-dimensional Euclidean distance transformation that determines at each point the minimal distance from the articular surface to the bone-cartilage interface [46–51].

More important than accuracy is the ability to distinguish changes of cartilage volume and thickness over time, which is determined by the reproducibility of the technique. There is significant disagreement in the literature as to the reproducibility of quantitative MR imaging–derived measurements of cartilage loss in the tibiofemoral compartments. Coefficients of variation for repeated measurements of total cartilage volume derived from standard three-dimensional SPGR sequences ranged between 1.8% and 8.2% [9,15,46,48,60,70,71], and in one study even 10% and 15% [58].

Wluka et al [72] reported that the annual rate of total tibial cartilage loss in a longitudinal study in OA patients amounted to 5.3% ± 5.2% (mean ± 1 SD) (95% confidence interval [CI] 4.4%–6.2%) per year, a value only slightly above most of the published reproducibility errors. The annual percentages of loss of medial and lateral tibial cartilage were 4.7% ± 6.5% (95% CI 3.6%–5.9%) and 5.3% ± 7.2% (95% CI 4.1%–6.6%), respectively [72]. Gandy et al [22] were not able to find any discernable change in cartilage volume in a cohort of OA patients followed with MR imaging over a 3-year period. Remarkably, radiologists' visual readings showed progression of cartilage loss in the same patients [73]. Difficulties in cartilage segmentation caused by low cartilage contrast in the three-dimensional SPGR sequence seemed to be largely responsible for the problems noted with quantitative cartilage measurements in that study [22].

Hardy et al [60] showed that the spatial resolution of the imaging sequence is of critical importance for reducing partial volume artifacts in cartilage MR imaging and for improving the reproducibility of quantitative measurements of cartilage loss. Changing the slice thickness from 1 to 0.5 mm resulted on average in a 2% decrease in coefficients of variation in the tibiofemoral compartments [60]. Similarly, a change in in-plane resolution from 0.55 to 0.275 mm caused a threefold decrease in coefficients of variation of repeated cartilage volume measurements. Alongside the high variability in published reproducibility errors [9,15,46,48,60,70,71] and the difficulties encountered by some investigators in segmenting the articular cartilage in OA patients [22,73], the results of Hardy et al [60] emphasize the need for novel three-dimensional imaging techniques with high contrast and high spatial resolution.

Image artifacts

In a recent study [19], the presence and severity of image artifacts on conventional two-dimensional FSE and three-dimensional SPGR sequences was evaluated for cartilage imaging. Four normal volunteers and 28 patients with OA of the knee (Kellgren-

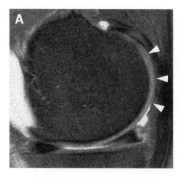
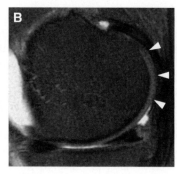
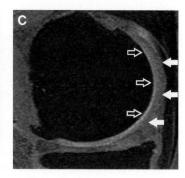

Fig. 3. Ambiguity of cartilage surface contour on three-dimensional SPGR image. MR images in a 60-year-old man. Sagittal short TE FSE image (4000/13) (*A*) and long TE FSE image (4000/39) with fat suppression (*B*) showing clearly defined cartilage contour (*arrowheads*) in the region of the posterior femoral condyle. (*C*) Sagittal fat-suppressed three-dimensional SPGR image (60/5, 40°-flip angle) shows ambiguous surface contour (*arrows*) on the posterior region of the femoral condyle cartilage. (*From* Yoshioka H, Stevens K, Genovese M, et al. Articular cartilage of knee: normal patterns at MR imaging that mimic disease in healthy subjects and patients with osteoarthritis. Radiology 2004;231:31–8; with permission.)

Lawrence grades I and II [74]) were prospectively studied with MR imaging (standard, conventional two-dimensional FSE [short TE and moderate TE], and three-dimensional SPGR). Imaging artifacts were noted. Signal intensities of cartilage, meniscus, joint capsule, synovial fluid, and muscle were measured and the tissue contrast was determined for each sequence [19].

The following artifacts were observed on FSE short TE, FSE long TE, and SPGR images, respectively: (1) ambiguity of the surface contour in the posterior region of the femoral condyle cartilage 0%, 0%, and 71.4% (Fig. 3); (2) linear high signal intensity in the deep zone adjacent to the subchondral bone of the femoral condyle 0%, 0%, and 92.9%; (3) pseudolaminar appearance in the posterior region of the femoral condyle cartilage 25%, 32.1%, and 85.7%; (4) truncation artifact in the patellofemoral compartment 25%, 21.4%, and 96.2%; (5) suscepti-

bility artifact on the cartilage surface caused by air or metal 10.7%, 10.7%, and 39.3%; (6) decreased signal intensity in the distal trochlear cartilage 100%, 100%, and 100%; (7) cartilage thinning in the central portion of the lateral femoral condyle adjacent to the anterior horn of the lateral meniscus 67.9%, 67.9%, and 75%; and (8) focal cartilage flattening in the posterior region of the femoral condyle 57.1%, 57.1%, and 32.1%. Limited contrast between cartilage and surrounding tissues resulted frequently in poor delineation of cartilage defects on conventional three-dimensional SPGR sequences (Fig. 4). Cartilage-meniscus contrast and cartilage–synovial fluid contrast were significantly greater on fat-suppressed two-dimensional FSE than on fat-suppressed three-dimensional SPGR images ($P < .001$) (Fig. 5).

Clearly, three-dimensional SPGR is hampered by multiple image artifacts that can obscure cartilage defects or artificially create defects [19]. These

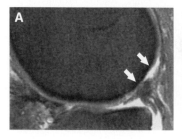
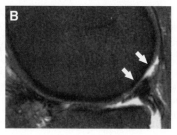
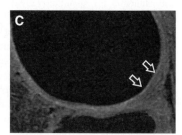

Fig. 4. Poor delineation of cartilage defect on three-dimensional SPGR sequence. MR images in the same patient seen in Fig. 2. Sagittal short TE FSE image (*A*) and long TE FSE image with fat suppression (*B*) show large, and in some areas full-thickness, cartilage defect (*arrows*) in the central and posterior regions of the lateral femoral condyle. (*C*) Although sagittal fat-suppressed three-dimensional SPGR image can demonstrate the cartilage defect, the margins are not well defined, particularly posteriorly (*open arrows*). The full-thickness cartilage defect was confirmed at arthroscopy. (*From* Yoshioka H, Stevens K, Genovese M, et al. Articular cartilage of knee: normal patterns at MR imaging that mimic disease in healthy subjects and patients with osteoarthritis. Radiology 2004;231:31–8; with permission.)

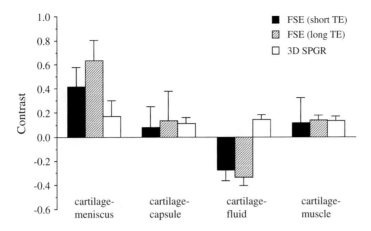

Fig. 5. Mean contrast between cartilage and adjacent structures for fat-suppressed FSE (short and long TE) images and fat-suppressed three-dimensional SPGR images. (*From* Yoshioka H, Stevens K, Genovese M, et al. Articular cartilage of knee: normal patterns at MR imaging that mimic disease in healthy subjects and patients with osteoarthritis. Radiology 2004;231: 31–8; with permission.)

artifacts along with the poor cartilage-meniscus and cartilage–synovial fluid contrast are the main reasons for the failure of automated software tools in segmenting cartilage for three-dimensional volume and thickness assessments that has been recently reported in OA patients [22]. The low prevalence of image artifacts and the substantially greater cartilage contrast afforded by two-dimensional FSE sequences suggest that a three-dimensional FSE imaging approach may have the potential to be relatively artifact free and to yield high image contrast and spatial resolution, essential prerequisites for accurate and reproducible visual and quantitative analysis of cartilage MR imaging.

Novel three-dimensional MR imaging pulse sequences

Three-dimensional driven equilibrium Fourier transform

A promising approach for imaging the patient with articular disorders is the DEFT technique [35–39]. Results of recent studies have shown that DEFT imaging provides contrast between cartilage and joint fluid by enhancing the signal from joint fluid, rather than by suppressing the signal from cartilage, as is the case with some sequences (Fig. 6).

DEFT produces image contrast that is a function of proton density, T1-T2, and TE-TR. The DEFT technique has been studied for many years, but so far has not been widely used. This is because T1 and T2 contrast compete in many tissues, so the T1-T2 DEFT

contrast tends to be flat. Fortunately, DEFT contrast is very well suited to imaging articular cartilage. Synovial fluid is high in signal intensity, and articular cartilage is intermediate in signal intensity. Bone appears low in signal intensity, and lipids are suppressed using a fat-saturation pulse. Hence, cartilage is easily distinguished from all of the adjacent tissues based on signal intensity alone, which greatly aids in segmentation and subsequent volume calculations.

Basic DEFT sequence includes a conventional spin echo pulse sequence followed by an additional refocusing pulse to form another echo, and then a reversed, excitation pulse to return any residual magnetization to the + z axis. This preserves the magnetization of the tissues that have longer T2, such as synovial fluid [35–39]. Three-dimensional DEFT

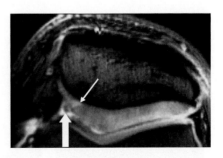

Fig. 6. Axial two-dimensional DEFT image through the patellofemoral articulation. The articular cartilage is well visualized. Joint fluid has high signal intensity. The thin arrow indicates a small cartilage fissure. The thick arrow indicates a cartilage flap arising from the fissure and extending into the fluid.

images with echo-planar readouts provide volumetric coverage in a clinically reasonable scan time and suggest that this sequence may be useful in clinical application (Fig. 7) [39].

The contrast-to-noise ratio efficiency between synovial fluid and cartilage has high values on fat-suppressed three-dimensional DEFT images. Fat-suppressed three-dimensional DEFT imaging is sensitive to signal intensity changes in degenerative cartilage.

Fat-suppressed three-dimensional DEFT imaging has several limitations. First, motion artifact is often seen, because of the longer scanning time than two-dimensional images and echo-planar imaging ghosts are seen because of the use of the echo-planar technique [35–39]. Second, for smaller field of view, stronger gradients are needed. Third, insufficient fat-suppression is often seen on fat-suppressed three-dimensional DEFT imaging, so further optimization may be necessary [39]. Finally, the areas of bone marrow edema are demonstrated as low or iso-signal intensity on fat-suppressed three-dimensional DEFT images, leading to underestimation of the extent of bone marrow edema [39]. Bone marrow edema is an important factor in diagnosis of OA. The complexity of signal intensity of bone marrow edema on fat-suppressed three-dimensional DEFT images seems to be caused by the fact that the contrast of DEFT imaging is dependent on the ratio T2-T1, not T1 or T2 relaxation times [39].

Water-excitation three-dimensional fast low-angle shot

Water-excitation three-dimensional fast low-angle shot is a sequence for faster imaging of articular cartilage defects, compared with conventional fat saturation three-dimensional fast low-angle shot [75]. The principle of water-excitation sequences is the selective excitation of non–fat-bound protons. Time-consuming spectral fat saturation to eliminate fat signal is not necessary. This provides significant advantage of a reduced acquisition time and additionally obviates chemical-shift artifacts. The water-excitation three-dimensional fast low-angle shot sequence has been recently described to provide image quality comparable with fat saturation three-dimensional fast low-angle shot [75].

Three-dimensional steady-state free precession and multipoint fat-water separation

SSFP imaging is an efficient, high-signal method for obtaining three-dimensional MR images [41]. The SSFP sequence is a rapid gradient-echo MR imaging technique that can yield a superior SNR compared with other gradient-echo techniques (Fig. 8).

The three-dimensional SSFP sequence is a fully balanced steady-state coherent imaging pulse sequence designed to produce high SNR images at very short sequence times (TR). The pulse sequence uses fully balanced gradients to rephase the transverse magnetization at the end of each TR interval. To achieve fat saturation in a steady state, it is important to bring the magnetization back to the steady state as quickly as possible to avoid artifacts. A half-alpha technique is used to store magnetization and then return it to steady state relatively quickly. This is repeated throughout the sequence at regular intervals. The use of multiecho fat-water separation allows the use of small TE increments, and can be

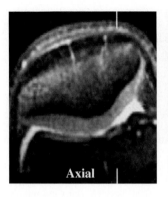 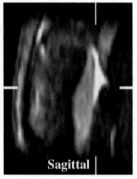 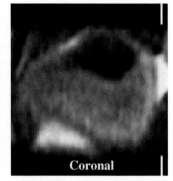

Fig. 7. Near isotropic three-dimensional DEFT acquisition using interleaved echoplanar acquisition. These images were obtained with the resolution of 0.5 × 0.5 × 1 mm, a field of view of 13 × 13 × 13 cm, and acquisition time of 11 minutes. The patellofemoral cartilage is demonstrated in the axial, sagittal, and coronal planes.

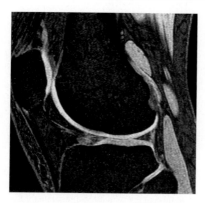

Fig. 8. Sagittal three-dimensional SSFP sequence obtained to the lateral tibiofemoral compartment (TR/TE/TI in milliseconds 6.6/1.2/20, flip angle 10°, 128 slices, 0.8-mm slice thickness, two excitations, matrix 256 × 256, field of view 14 cm, acquisition time 8 minutes). The sequence provides near isotropic resolution. Cartilage signal is very high even compared with the surrounding tissues resulting in excellent contrast-to-noise ratio.

applied to three-dimensional SSFP imaging of articular cartilage imaging.

SSFP sequence has excellent contrast behavior with varying dependence on T1 and T2. Many different methods of SSFP-based imaging are available for imaging the articular cartilage, and all have higher cartilage signal compared with conventional techniques. Synovial fluid appears bright on SSFP images, because of its long T2.

With recent advances in MR imaging gradient hardware, it is now possible to use SSFP imaging without the limitations of banding or off-resonance artifacts. The best immunity to off-resonance artifacts when using SSFP [41], is a short repetition time (<5 milliseconds). The major limitation of SSFP is image degradation caused by local magnetic field inhomogeneities if the TR is long.

Another similar approach that may provide more reliable fat suppression is Dixon SSFP imaging. This technique is faster and can provide more reliable fat suppression than fat-suppressed three-dimensional SPGR imaging. Dixon SSFP imaging may be especially useful at high field strength.

A final steady-state method of cartilage imaging that is useful at 3 T is SSFP imaging with intermittent fat saturation. This technique preserves the SNR advantage of SSFP imaging, but provides uniform fat saturation, even at high field strength. The overall SNR efficiency and speed of the SSFP-based techniques make them very attractive for cartilage imaging.

Three-dimensional fluctuating equilibrium MR imaging

Fluctuating equilibrium MR imaging [76,77] is a variant of SSFP imaging that may be useful in imaging cartilage in the knee [76]. In fluctuating equilibrium MR imaging, each phase-encoding step is repeated twice, once with a 90x and once with a 90y. The k-space data are then parsed to form two complete data sets, which are reconstructed into fat and water images. With a repetition time of 6.6 milliseconds, high-resolution three-dimensional imaging of cartilage is possible in about 2 minutes.

As with DEFT imaging, fluctuating equilibrium MR imaging produces contrast based on the ratio of T1-T2 in tissues. This results in bright fluid signal while preserving cartilage signal. The largest disadvantage of fluctuating equilibrium MR imaging and SSFP techniques is sensitivity to off-resonance artifacts [76].

Three-dimensional fast spin echo imaging

Standard two-dimensional FSE and three-dimensional SPGR sequences yield anisotropic voxels. Anisotropic voxels are difficult to process and can result in artifacts. Specifically, the poor slice profile, unwanted magnetization transfer effects, high-power deposition, or poor SNR of conventional two-dimensional slice-selective and multislab three-dimensional MR imaging methods can cause problems. A single-slab three-dimensional FSE sequence has been developed by the University of Virginia at Arlington and Brigham and Women's Hospital for neuroimaging [20,78] and has been tested to obtain high-resolution images of the brain [20] and petrous bone [21]. Images in these studies have near-isotropic voxel sizes of $0.4 \times 0.4 \times 0.6$ mm^3, which allows one to use isotropic image processing techniques without introducing significant artifacts.

The novel three-dimensional FSE sequence uses a single-slab technique with hard pulses for the excitation and for the refocusing radiofrequency pulses. This results in very small effective echo time and echo spacing. The novel three-dimensional FSE sequence differs from the product sequences of major MR imaging companies in many aspects. First, it has single-slab coverage for the whole field of view, whereas many product sequences are multislab because of the long sequence time needed. Single-slab three-dimensional MR imaging gives high SNR and eliminates unwanted magnetization transfer and boundary image artifacts. Second, the

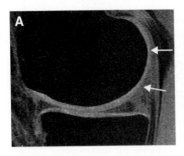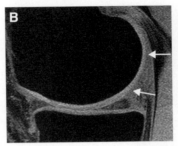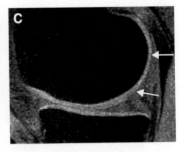

Fig. 9. Comparison of three-dimensional SPGR, three-dimensional SS-SPGR (water selective excitation gradient echo), and three-dimensional FSE images. Three-dimensional SPGR (*A*), three-dimensional SS-SPGR (*B*), and three-dimensional FSE images (*C*). Contrast between cartilage and posterior capsule (*arrows*) is best in this patient with three-dimensional FSE. This is a problem area for automated or semiautomated segmentation of cartilage for subsequent quantitative analysis, such as measurements of volume and thickness. Blurring is not a problem with this sequence because of its short echo times.

sequence uses hard excitation and refocusing radio-frequency pulses that can significantly reduce echo spacing compared with the conventional soft pulse approach. This greatly helps to reduce the effective echo time in T1-weighted imaging and helps to reduce the echo train length that shortens the total scan time and eliminates image artifacts caused by signal quick decaying. Third, dynamic receiver gains can be implemented during three-dimensional data acquisition, which improves SNR compared with the conventional single gain approach. A twofold to threefold SNR gain can be obtained with this approach [20].

Three-dimensional FSE provides image contrast similar to the one experienced with routinely used two-dimensional FSE sequences (Fig. 9). The sequence can provide isotropic image resolution. In the future, this may create the opportunity of single-

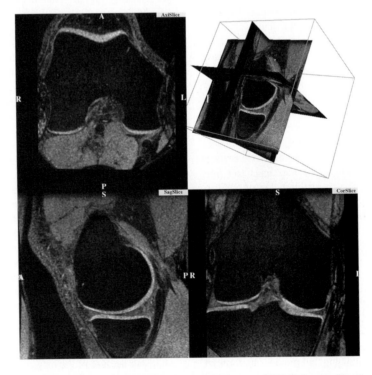

Fig. 10. Simultaneous multiplanar display of three-dimensional FSE images, 3-T MR image. The three-dimensional FSE sequence is acquired with near isotropic resolution (0.47 × 0.6 × 0.5 mm). Images can be displayed in any desired plane without apparent loss in image quality. Partial volume artifacts are greatly reduced using this approach.

pulse sequence knee MR imaging with isotropic resolution; in this setting, the radiologist can rotate and review the acquisition in any desired orientation without loss of spatial resolution (Fig. 10).

Acknowledgments

The authors would like to thank Inez Wu for assistance in preparing the manuscript.

References

[1] Doherty M, Hutton C, Bayliss MT. Osteoarthritis. In: Maddison PJ, et al, editors. Oxford textbook of rheumatology. Oxford: Oxford University Press; 1993. p. 959–83.

[2] Brittberg M, Lindahl A, Nilsson A, et al. Treatment of deep cartilage defects in the knee with autologous chondrocyte transplantation. N Engl J Med 1994; 331:889–95.

[3] Brittberg M, Lindahl A, Homminga G, et al. A critical analysis of cartilage repair. Acta Orthop Scand 1997;68:186–91.

[4] Garrett JC. Osteochondral allografts for reconstruction of articular defects of the knee. Instr Course Lect 1998; 47:517–22.

[5] Stevenson S, Dannucci GA, Sharkey NA, et al. The fate of articular cartilage after transplantation of fresh and cryopreserved tissue-antigen-matched and mismatched osteochondral allografts in dogs. J Bone Joint Surg 1989;71:1297–307.

[6] Bobic V. Arthroscopic osteochondral autograft transplantation in anterior cruciate ligament reconstruction: a preliminary clinical study. Knee Surg Sports Traumatol Arthrosc 1996;3:262–4.

[7] Buckland-Wright JC, Macfarlane DG, Williams SA, et al. Accuracy and precision of joint space width measurements in standard and macroradiographs of osteoarthritic knees. Ann Rheum Dis 1995;54:872–80.

[8] Buckland-Wright JC, Wolfe F, Ward RJ, et al. Substantial superiority of semiflexed (MTP) views in knee osteoarthritis: a comparative radiographic study, without fluoroscopy, of standing extended, semiflexed (MTP), and schuss views. J Rheumatol 1999;26: 2664–74.

[9] Burgkart R, Glaser C, Hyhlik-Durr A, et al. Magnetic resonance imaging-based assessment of cartilage loss in severe osteoarthritis: accuracy, precision, and diagnostic value. Arthritis Rheum 2001;44:2072–7.

[10] Disler DG, McCauley TR, Wirth CR, et al. Detection of knee hyaline cartilage defects using fat-suppressed three-dimensional spoiled gradient-echo MR imaging: comparison with standard MR imaging and correlation with arthroscopy. AJR Am J Roentgenol 1995;165: 377–82.

[11] Disler DG, McCauley TR, Kelman CG, et al. Fat-suppressed three-dimensional spoiled gradient-echo MR imaging of hyaline cartilage defects in the knee: comparison with standard MR imaging and arthroscopy. AJR Am J Roentgenol 1996;167:127–32.

[12] Disler DG. Fat-suppressed three-dimensional spoiled gradient recalled MR imaging: assessment of articular and physeal hyaline cartilage. AJR Am J Roentgenol 1997;169:1117–23.

[13] Potter HG, Linklater JM, Allen AA, et al. Magnetic resonance imaging of articular cartilage in the knee: an evaluation with use of fast-spin-echo imaging. J Bone Joint Surg Am 1998;80:1276–84.

[14] McCauley TR, Kier R, Lynch KJ, et al. Chondromalacia patellae: diagnosis with MR imaging. AJR Am J Roentgenol 1992;158:101–5.

[15] Peterfy CG, van Dijke CF, Janzen DL, et al. Quantification of articular cartilage in the knee with pulsed saturation transfer subtraction and fat-suppressed MR imaging: optimization and validation. Radiology 1994; 192:485–91.

[16] Peterfy CG, Majumdar S, Lang P, et al. MR imaging of the arthritic knee: improved discrimination of cartilage, synovium, and effusion with pulsed saturation transfer and fat-suppressed T1-weighted sequences. Radiology 1994;191:413–9.

[17] Broderick LS, Turner DA, Renfrew DL, et al. Severity of articular cartilage abnormality in patients with osteoarthritis: evaluation with fast spin-echo MR vs arthroscopy. AJR Am J Roentgenol 1994;162:99–103.

[18] Recht MP, Piraino DW, Paletta GA, et al. Accuracy of fat-suppressed three-dimensional spoiled gradient-echo FLASH MR imaging in the detection of patellofemoral articular cartilage abnormalities. Radiology 1996;198: 209–12.

[19] Yoshioka H, Stevens K, Genovese M, et al. Articular cartilage of knee: normal patterns at MR imaging that mimic disease in healthy subjects and patients with osteoarthritis. Radiology 2004;231:31–8.

[20] Zhao L, et al. A high-resolution clinical whole-brain scan using single-slab three-dimensional T1W, T2W and FLAIR fast spin-echo sequences. In: Proceedings of the International Society of Magnetic Resonance in Medicine. Honolulu, Hawaii, May, 2002. p. 1294.

[21] Zhao L, Bartling S, Mugler J, et al. High-resolution MR imaging of the petrous bone using a single-slab three-dimensional T2-weighted fast spin-echo sequence. In: Proceedings of the International Society of Magnetic Resonance in Medicine. Honolulu, Hawaii, May, 2002. p. 312.

[22] Gandy SJ, Dieppe PA, Keen MC, et al. No loss of cartilage volume over three years in patients with knee osteoarthritis as assessed by magnetic resonance imaging. Osteoarthritis Cartilage 2002;10:929–37.

[23] Conway WF, Hayes CW, Loughran T, et al. Cross-sectional imaging of the patellofemoral joint and surrounding structures. Radiographics 1991;11:195–217.

[24] Hayes C, Sawyer R, Conway W. Patellar cartilage lesions: in vitro detection and staging with MR imag-

ing and pathologic correlation. Radiology 1990;176: 763–6.

[25] Karvonen RL, Negendank WG, Fraser SM, et al. Articular cartilage defects of the knee: correlation between magnetic resonance imaging and gross pathology. Ann Rheum Dis 1990;49:672–5.

[26] Recht MP, Resnick D. MR imaging of articular cartilage: current status and future directions. AJR Am J Roentgenol 1994;163:283–90.

[27] Recht MP, Kramer J, Marcelis S, et al. Abnormalities of articular cartilage in the knee: analysis of available MR techniques. Radiology 1993;187:473–8.

[28] Yulish BS, Montanez J, Goodfellow DB, et al. Chondromalacia patellae: assessment with MR imaging. Radiology 1987;164:763–6.

[29] Lehner KB, et al. Structure, function, and degeneration of bovine hyaline cartilage: assessment with MR imaging in vitro. Radiology 1989;170:495–9.

[30] Henkelman RM, Stanisz GJ, Kim JK, et al. Anisotropy of NMR properties of tissues. Magn Reson Med 1994; 32:592–601.

[31] Woolf SD, et al. Magnetization transfer contrast: MR imaging of the knee. Radiology 1991;179:623–8.

[32] Pauly JM, et al. Slice selective excitation for very short T2 species. In: Society for Magnetic Resonance in Medicine: book of abstracts; 1990.

[33] Brossmann J, et al. Short echo time projection reconstruction MR imaging of cartilage: comparison with fat-suppressed spoiled GRASS and magnetization transfer contrast MR imaging. Radiology 1997;203:501–7.

[34] Gold G, Pauly JM, Macovski A, et al. MR spectroscopic imaging of collagen: tendons and knee menisci. Magn Reson Med 1995;34:647–54.

[35] Hargreaves BA, et al. Comparison of novel sequences for imaging articular cartilage. In: Proceedings of the International Society of Magnetic Resonance in Medicine; 2002.

[36] Hargreaves BA, et al. Technical considerations for DEFT imaging. In: International Society for Magnetic Resonance in Medicine. Sydney, Australia; 1998.

[37] Hargreaves BA, et al. Imaging of articular cartilage using driven equilibrium. In: International Society for Magnetic Resonance in Medicine. Sydney, Australia; 1998.

[38] Hargreaves BA, Gold GE, Lang PK, et al. MR imaging of articular cartilage using driven equilibrium. Magn Reson Med 1999;42:695–703.

[39] Yoshioka H, Stevens K, Hargreaves BA, et al. Magnetic resonance imaging of articular cartilage of the knee: comparison between fat-suppressed three-dimensional SPGR imaging, fat-suppressed FSE imaging, and fat-suppressed three-dimensional DEFT imaging, and correlation with arthroscopy. J Magn Reson Imaging 2004;20:857–64.

[40] Yoshioka H, Alley M, Steines D, et al. Imaging of the articular cartilage in osteoarthritis of the knee joint: 3D spatial-spectral spoiled gradient-echo versus fat-suppressed 3D spoiled gradient-echo MR imaging. J Magn Reson Imaging 2003;18:66–71.

[41] Reeder SB, Pelc NJ, Alley MT, et al. Rapid MR imaging of articular cartilage with steady-state free precession and multipoint fat-water separation. AJR Am J Roentgenol 2003;180:357–62.

[42] Lang P, et al. Cartilage MR imaging at 3.0T: comparison of 3D SPGR, 3D MFAST and 3D FSE sequences. In: OARSI Omeract Workshop for Consensus in Osteoarthritis Imaging. Bethesda, Maryland; 2002.

[43] Bredella MA, Tirman PF, Peterfy CG, et al. Accuracy of T2-weighted fast spin-echo MR imaging with fat saturation in detecting cartilage defects in the knee: comparison with arthroscopy in 130 patients. AJR Am J Roentgenol 1999;172:1073–80.

[44] Kawahara Y, Uetani M, Nakahara N, et al. Fast spin-echo MR of the articular cartilage in the osteoarthrotic knee: correlation of MR and arthroscopic findings. Acta Radiol 1998;39:120–5.

[45] McNicholas MJ, Brooksbank AJ, Walker CM. Observer agreement analysis of MRI grading of knee osteoarthritis. J R Coll Surg Edinb 1999;167:757–60.

[46] Eckstein F, Gavazzeni A, Sittek H, et al. Determination of knee joint cartilage thickness using three-dimensional magnetic resonance chondro-crassometry (3D MR-CCM). Magn Reson Med 1996;36:256–65.

[47] Eckstein F, Sittek H, Gavazzeni A, et al. Magnetic resonance chondro-crassometry (MR CCM): a method for accurate determination of articular cartilage thickness? Magn Reson Med 1996;35:89–96.

[48] Eckstein F, Westhoff J, Sittek H, et al. In vivo reproducibility of three-dimensional cartilage volume and thickness measurements with MR imaging. AJR Am J Roentgenol 1998;170:593–7.

[49] Stammberger T, Hohe J, Englmeier KH, et al. Patellofemoral joint cartilage thickness and contact areas from MRI in patients with osteoarthritis. In: International Society for Magnetic Resonance in Medicine. Sydney, Australia; 1998.

[50] Stammberger T, Eckstein F, Englmeier KH, et al. Determination of 3D cartilage thickness data from MR imaging: computational method and reproducibility in the living. Magn Reson Med 1999;41:529–36.

[51] Tieschky M, Faber S, Haubner M, et al. Repeatability of patellar cartilage thickness patterns in the living, using a fat-suppressed magnetic resonance imaging sequence with short acquisition time and three-dimensional data processing. J Orthop Res 1997;15:808–13.

[52] Peterfy C, van Dijke CF, Lu Y, et al. Quantification of the volume of articular cartilage in the metacarpophalangeal joints of the hand: accuracy and precision of three-dimensional MR imaging. AJR Am J Roentgenol 1995;165:371–5.

[53] Piplani MA, Disler DG, McCauley TR, et al. Articular cartilage volume in the knee: semiautomated determination from three-dimensional reformations of MR images. Radiology 1996;198:855–9.

[54] Steines D, Napel S, Lang P. Measuring volume of articular cartilage defects in osteoarthritis using MRI: validation of a new method. In: Radiological Society of North America. Chicago, Illinois, December, 2000.

[55] Steines D, Napel S, Lang P, et al. Segmentation of osteoarthritic femoral cartilage using live wire. Presented at the ISMRM Eighth Scientific Meeting. Denver, Colorado, 2000.

[56] Stammberger T, Eckstein F, Michaelis M, et al. Interobserver reproducibility of quantitative cartilage measurements: comparison of B-spline snakes and manual segmentation. Magn Reson Imaging 1999;17: 1033–42.

[57] Warfield S, Winalski C, Jolesz F, et al. Automatic segmentation of MRI of the knee. In: International Society for Magnetic Resonance in Medicine. Sydney, Australia, April, 1998. p. 563.

[58] Pilch L, Stewart C, Gordon D, et al. Assessment of cartilage volume in the femorotibial joint with magnetic resonance imaging and 3D computer reconstruction. J Rheumatol 1994;21:2307–21.

[59] Hardy PA, Nammalwar P, Kuo S. Measuring the thickness of articular cartilage from MR images. J Magn Reson Imaging 2001;13:120–6.

[60] Hardya PA, Newmark R, Liu YM, et al. The influence of the resolution and contrast on measuring the articular cartilage volume in magnetic resonance images. Magn Reson Imaging 2000;18:965–72.

[61] Westin CF, Wigstrom L, Loock T, et al. Three-dimensional adaptive filtering in magnetic resonance angiography. J Magn Reson Imaging 2001;14:63–71.

[62] Westin CF, Richolt J, Moharir V, et al. Affine adaptive filtering of CT data. Med Image Anal 2000;4:161–77.

[63] Rodrigues-Florido MA, Krissian K, Ruiz-Alzola J, et al. Comparison of two restoration techniques in the context of 3D medical imaging. In: MICCAI 2000: Fourth International Conference on Medical Image Computing and Computer-Assisted Intervention. Utrecht, The Netherlands: Springer Verlag; 2000.

[64] Sijbers J, Scheunders P, Verhoye M, et al. Watershed-based segmentation of 3D MR data for volume quantization. Magn Reson Imaging 1997;15:679–88.

[65] Falcão AX, Udupa JK. Segmentation of 3D objects using live wire. In: SPIE Medical Imaging. Newport Beach, California. 1997;3024:25.

[66] Falcão AX, Udupa JK. User-steered image segmentation paradigms: live wire and live lane. GMIP 1998; 60:233–60.

[67] Warfield SK, Kaus M, Jolesz FA, et al. Adaptive, template moderated, spatially varying statistical classification. Med Image Anal 2000;4:43–55.

[68] Warfield S, Dengler J, Zaers J, et al. Automatic identification of gray matter structures from MRI to improve the segmentation of white matter lesions. J Image Guid Surg 1995;1:326–38.

[69] Iosifescu DV, Shenton ME, Warfield SK, et al. An automated registration algorithm for measuring MRI subcortical brain structures. Neuroimage 1997; 6:13–25.

[70] Eckstein F, Heudorfer L, Faber SC, et al. Long-term and resegmentation precision of quantitative cartilage MR imaging (qMRI). Osteoarthritis Cartilage 2002;10: 922–8.

[71] Glaser C, Faber S, Eckstein F, et al. Optimization and validation of a rapid high-resolution T1-w 3D FLASH water excitation MRI sequence for the quantitative assessment of articular cartilage volume and thickness. Magn Reson Imaging 2001;19:177–85.

[72] Wluka AE, Stuckey S, Snaddon J, et al. The determinants of change in tibial cartilage volume in osteoarthritic knees. Arthritis Rheum 2002;46: 2065–72.

[73] Gandy SJ, Brett AD, Pirppe PA, et al. No apparent progressive change to knee cartilage volumes over one year in rheumatoid and osteoarthritis. In: Proceedings of the International Society of Magnetic Resonance in Medicine. Denver, Colorado, (CO); 2000.

[74] Kellgren J, Lawrence J. Radiological assessment of osteoarthritis. Ann Rheum Dis 1957;16:494–501.

[75] Hyhlik-Durr A, Faber S, Burgkart R, et al. Precision of tibial cartilage morphometry with a coronal water-excitation MR sequence. Eur Radiol 2000;10: 297–303.

[76] Vasnawala SS, Pauly JM, Nishimura DG, et al. MR imaging of knee cartilage with FEMR. Skeletal Radiol 2002;31:574–80.

[77] Vasanawala SS, Pauly JM, Nishimura DG. Fluctuating equilibrium MRI. Magn Reson Med 1999;42: 876–83.

[78] Mugler III JP, Bao S, Mulkern RV, et al. Optimized single-slab three-dimensional spin-echo MR imaging of the brain. Radiology 2000;216:891–9.

ELSEVIER
SAUNDERS

Radiol Clin N Am 43 (2005) 641 – 653

RADIOLOGIC
CLINICS
of North America

New Techniques for Cartilage Imaging: T2 Relaxation Time and Diffusion-Weighted MR Imaging

Christian Glaser, MD*

Musculoskeletal Imaging, Division of General Radiography, Department of Clinical Radiology,
Ludwig-Maximilians-Universität München, Munich, Germany

Integrity of the articular cartilage is a decisive prerequisite for adequate long-term joint function [1,2]. Damage to the articular cartilage is generally acknowledged as an early factor in the process of irreversible joint degeneration (ie, osteoarthritis [OA]), as a major and global socioeconomic burden [3–5]. Efforts are being made to develop and further refine cartilage-dedicated therapeutic strategies targeted to the therapy of early stages of OA [6]. The aim is to reconstitute a congruous and long-lasting biomechanically valid joint surface, and those techniques in part directly affect the internal matrix of the cartilage. Examples are osteochondral grafting, regrowth of cartilaginous matrix using chondrocyte transplantation techniques, or inducing growth of repair tissue by microfracture or drilling techniques. Another techniques is osteotomy. More recently, efforts are being made to develop drugs with the long-term goal to achieve structure modifying effects.

These efforts create a strong need for a noninvasive diagnostic tool that can be applied to a high number of patients. It should be able to give a valid estimate of the status of the cartilage by reliably discriminating intact cartilage from various grades of damaged cartilage. In view of the evaluation of the efficacy of new treatment concepts it should provide quantitative data to facilitate follow-up studies implying knowledge on reproducibility of the respective technique. Moreover, such techniques should be able to differentiate between potentially reversible and irreversible stages of cartilage alterations.

Current dedicated and mostly quantitative MR imaging–based diagnostic approaches to the articular cartilage include three-dimensional volumetric assessment, gadolinium-enhanced proteoglycan imaging, magnetization-transfer imaging, T2 relaxation time quantitation, and diffusion-weighted imaging. Among them the most clinical experience, validation, and reproducibility data are available for three-dimensional assessment of cartilage volume, thickness, and joint surface area. It is probably the easiest parameter to assess profiting from a uniform, high signal intensity depiction of cartilage in three-dimensional fat-suppressed T1-weighted gradient echo sequences. Volumetric assessment is followed by the gadolinium-enhanced proteoglycan imaging technique applying the negatively charged ion gadolinium with its high affinity to areas of proteoglycan depletion in the cartilage. It targets alterations of proteoglycan content and distribution in the cartilage, a process earlier involved in cartilage degeneration than open substance loss. Less is known about T2 relaxation time assessment and assessment of diffusion in cartilage. There has been an impressive increase of data available on the former technique in the last 5 years.

* Leiter Allgemeine Radiologie und Funktionsbereich Orthop. Bildgebung, Institut für Klinische Radiologie der Ludwig-Maximilians-Universität München, Campus Großhadern, Marchioninistraße 15, München 81377, Germany.

E-mail address: christian.glaser@med.uni-muenchen.de

Basic principles of cartilage morphology

Knowledge of the macroscopic and microscopic morphology of intact cartilage and of the alterations occurring in the cartilage matrix during the process of degeneration is mandatory to understand the MR imaging aspect of cartilage in the various techniques. In view of its morphology, the cartilage's MR imaging appearance is governed by three factors that are considered to be interdependent of each other: (1) its biochemical composition (in terms of amount and concentration of its constituents); (2) its degree of hydration; and (3) its internal architecture (in terms of orientation and spatial relationships of the constituents to each other).

The structure-giving component of cartilage is considered to be the collagenous fibers. These fibers show a very low biologic turnover and exhibit high tensile stiffness [1,7,8]. They are arranged in form of a network with various degrees of cross-linking. Their alignment relative to the surface or the uncal-cified cartilage-to-calcified cartilage interface (tide mark) is used to describe the various layers (zones) of the cartilage (Fig. 1). The most superficial layer is the tangential layer with a fiber alignment predominantly parallel (tangential) to the surface. It is followed by the transitional layer with no clearly predominant (isotropic) fiber orientation. Deep to the transitional layer is the radial layer with a predominant fiber alignment vertical (radial) to the tide mark [9,10]. Immediately above the tide mark a small area of isotropic fiber orientation may be seen. The relative height of the cartilage layers depends on the age, skeletal maturity, and species examined and varies according to the joint, the joint compartments, and the location within a joint compartment [11–13]. In this connection it is important to note that the anisotropy of the collagenous fibers is a three-

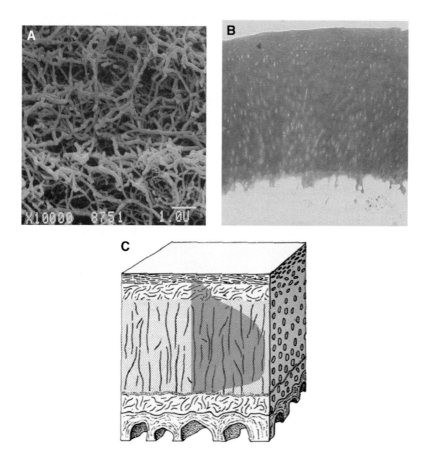

Fig. 1. Anatomy of articular cartilage. (*A*) Scanning electron micrograph of collagenous fibers showing a random alignment in the transitional layer of cartilage. The white bar represents 1 μm. (*B*) The pores of this network are occupied by hydrated proteoglycans. Their concentration increases from the surface down to the tide mark visualized qualitatively with histologic techniques, such as safranin O staining. (*C*) Schematic drawing of the zonal architecture, which is described by the predominant alignment of the collagenous fibers. This zonal architecture is overlayed by the bell-shaped distribution of proteoglycan concentration.

dimensional property of cartilage (ie, that in addition to the anisotropy along a radial [from surface to tide mark] axis used to define the separate layers, there may be anisotropy within the plane of a previously defined layer). An illustrative example is the tangential layer where anisotropy within this layer parallel to the surface could be shown using the split line technique in anatomic samples [14].

The spaces in between the collagenous fibers are filled by proteoglycans and water commonly referred to as a "gellike substance" in which the water seems to be attracted by the highly negatively charged side chains of the proteoglycans. This proteoglycan water gel is able to provide a swelling pressure to the cartilage [1]. Whether the spatial alignment of the proteoglycans is determined by the collagenous fiber architecture is not clear. Only the reciprocal stabilization of collagenous fibers and proteoglycan water gel, however, are able to achieve the unique mechanical properties of cartilage with its high ability to bear tensile and compressive stresses and strains and superficial wear. The concentration of the proteoglycans across the depth of the cartilage follows a bell-like shape that differs from the layers as defined by the collagenous fibers. It shows low values close to the surface that increase to a maximum at about 50% to 80% depth and then decrease when further approaching the tide mark [15]. The degree of hydration depends on the concentration of proteoglycans. It depends on the degree of unfolding of the proteoglycan side chains (ie, their degree of entrapment by and the integrity of the collagenous fibers).

Early changes of the cartilage matrix before open loss of cartilage substance occurs are reported to be alterations of the water content in the order of magnitude of 5% to 15% [16,17], a reduction of proteoglycans, and disintegration of the collagenous fiber network. The two former are potentially reversible, whereas the latter constitutes an irreversible alteration of the cartilage matrix followed by progressive cartilage damage and erosion. It is not clear that potentially reversible alterations and irreversible changes occur at distinctly definable time points in the course of OA. Presumably, there is overlap between them and features of all or several pathologic changes are present when assessment of cartilage is performed.

T2 relaxation time measurements: technical aspects

The T2 relaxation time is a tissue-specific time constant describing the decay of transverse magneti-zation of tissues. It is determined by measuring the MR imaging signal intensity at various echo times (S[t]) and by then fitting a corresponding exponential equation to these data points: $S(t) = S_0 \exp(-t/T2)$, from which the T2 relaxation time constant can then be calculated (Fig. 2).

In clinical applications, the typical MR imaging pulse sequence for T2 relaxation time quantitation is usually a Carr-Purcell-Meiboom-Gill (CPMG) multi-echo multislice sequence able to cover a complete joint in an acceptable imaging time of less than 15 minutes [18–27]. In clinical whole-body MR imagers the achievable minimal echo spacing is around 8 to 12 milliseconds [18–27] and accordingly 8 to 10 echoes can be implemented in the first 100 milliseconds after the excitation pulse. MR imaging signal from cartilage that is acquired after these initial 100 milliseconds at 1.5 T usually becomes very close to the level of background noise because of the short T2 relaxation times of the articular cartilage. The results of Duewell et al [28] suggest that T2 relaxation time of patellar cartilage is shorter by about 30% at 4 T as compared with 1.5 T in accordance with the observations of Mlynarik et al [29] that transverse relaxivity is increased in cartilage with increasing B_0 from 2.95 to 7.05 T. Gold et al [30] observed a 12% reduction of knee cartilage T2 relaxation times from 1.5 to 3.T in five healthy volunteers. To avoid saturation effects causing errors in the measurement of T2 relaxation time it may be beneficial to keep TR about five times the estimated T1 relaxation time of the tissue being analyzed (when > 95% of initial longitudinal magnetization has recovered) [31]. Because chemical shift artifacts may artificially increase apparent T2 relaxation times if their spatial extension along the frequency-encoding direction reaches the cartilage immediately adjacent to the subchondral bone [27], fat saturation may be beneficial.

The multiecho multislice sequence design implies a contribution of stimulated echoes to the calculated (apparent) T2 relaxation time [20,21,32] arising from slice-selective refocusing pulses with imprecise section profiles. Tipping transverse magnetization components by almost 180 degrees during refocusing results in a component of this (theoretically) purely transverse magnetization out of the transverse plane. A certain longitudinal component is created that is prone to T1 relaxation and increases with time. Parts of it are reconverted into transverse magnetization by following radiofrequency pulses in the multiecho sequence and interfere with the primary (ie, Hahn) echoes from the original transverse magnetization. This leads to an artificially increased apparent T2 relaxation time as compared with spectroscopic T2

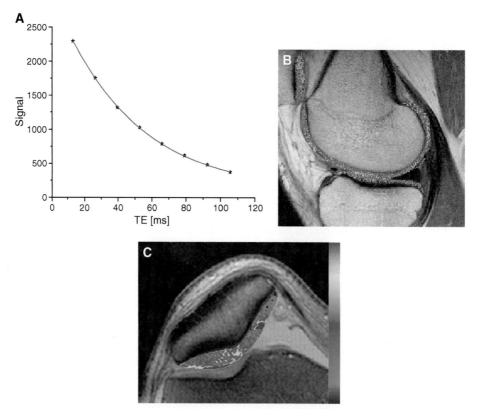

Fig. 2. T2 relaxation time mapping of articular cartilage. (*A*) Signal decay measured at subsequent echo time points {(repetition time 3000 ms)/[(echo time)$_{min}$ 12.8 ms]/(echo train length 8)} and fitted exponential T2 decay function. (*B*) T2 map of a healthy volunteer's knee cartilage in the sagittal plane. After segmentation of the cartilage and overlay of this segmentation on the T2 map, the T2 values have been projected on a corresponding section of a routine T1-weighted sequence for better anatomic orientation. (*C*) Patellar T2 map of a patient with anterior cruciate ligament tear subsequent to a sports injury. The focal area of strongly increased T2 corresponds to a patellar cartilage fissure in the medial patellar facet.

relaxation time measurements or single spin echo–based measurements. Technical means proposed to reduce these stimulated echoes are reduction of section-selective gradient strength; spoiler gradients with alternating amplitudes around (ie, bracketing) the section-selective gradients; composite pulses; phase cycling; and increase of radiofrequency refocusing pulse duration [20,21,32,33]. A reduced section-selective gradient strength improves the section profile but introduces magnetization transfer effects and slice-to-slice crosstalk, especially if the slices are acquired without gap, seriously instead of interleaved, and if the time spacing of data acquisition from adjacent slices is smaller than tissue's T1. Overall, introduction of such elements may even lead to an underestimation of apparent T2 relaxation time. Interestingly, the amount of overestimation or underestimation seems to differ between phantoms and

cartilage; after specific sequence optimization it can be reduced from up to 48% down to 5% to 15% [20,21]. At present, the individual sequence technique should carefully be taken into account when comparing T2 relaxation time measurements between different institutions. Although in clinical imaging the estimated contribution of diffusion effects to signal decay in single spin echo sequences is small [21] in cartilage, diffusional contribution in multiecho sequences is not yet clear.

Factors influencing T2 relaxation: magic angle effect

Experimental data [29,34–36] suggest different (rapid and slow) T2 relaxation components in cartilage. It is generally agreed on that using echo times and echo spacing in the order of magnitude of 8 to 12 milliseconds, only the slow T2 relaxation components can be assessed [37,38]. The dominant mechanism of T2 relaxation in cartilage at field strengths not above 3 T is considered to be dipolar interaction of water molecules [29]. Presumably, these water molecules may belong to two different pools [34] with T2 relaxation time constants of 20 and 55 milliseconds, respectively: one as free water, thought to be responsible for a less orientation-dependent relaxation component; the other as associated with macromolecules, thought to be responsible for a more orientation-dependent relaxation component [29,37].

Similar to tendons, T2 relaxation times show a strong dependence on the orientation of the cartilage (sample) to the main magnetic field [23,29,37, 39–41], known as the "magic angle" effect. According to theory [42], dipolar interaction becomes minimal between water molecules oriented at an angle of approximately 55 degrees to B_0 and transverse relaxation time is maximal leading to increased apparent MR imaging signal. This could be demonstrated for tendons as tissues containing a high amount of strongly ordered (predominantly type I) collagen-restricting water motion. The observed signal increase in tendons is described to be in the order of magnitude of up to a factor of 100 [43]. In clinical T2 relaxation time assessment such topographic variations of T2 relaxation times should be taken into account, especially when it is a question of discriminating healthy and diseased cartilage. The orientation dependence of T2 relaxation times might be a valuable tool for ultrastructural analysis of the cartilage matrix.

In ex vivo cartilage studies T2 relaxivity has been shown to be minimal (and hence T2 relaxation times to be maximal) when the alignment of the cartilage sample (ie, the normal to its surface) is also at an angle of approximately 55 degrees (the magic angle) to the main magnetic field [37,39]. The effect is smaller by several orders of magnitude compared with tendons. The ex vivo studies show a relative decrease by 30% to 80% of cartilage T2 relaxation times in various heights in the cartilage when comparing the magic angle position and angulation of the probes parallel or perpendicular to the main magnetic field [35,37,39,40,44]. This effect is most pronounced in the deeper regions thought to correspond to the radial layer; it is least pronounced in the region of the cartilage presumed to be the transitional layer. Chondrocyte accumulations are thought to cause local alterations of the course of the collagenous fibers and local alterations of the magic angle

conditions explaining unexpected local T2 inhomogeneities in the deeper zones of cartilage [45]. An in vivo observation [23] describes a relative increase (9%–30%) of T2 relaxation times at different locations (and hence different orientations of the cartilage to B_0) over the strongly curved femoral condyles to be more prominent in the superficial portions of the cartilage than in its deeper portions. These observations and the differences between some of the ex vivo and in vivo studies are a matter of debate [37,38,46,47]. Among the factors cited to explain them are partial volume averaging in clinical imaging and a nonuniform morphology of the cartilage matrix across a joint surface [11–13]. The similarity in location and extent of regions with distinct T2 relaxation times and alterations of T2 relaxation times with their orientation relative to B_0, and layers with a typical orientation of the collagenous fibers (in cartilage, mainly type II collagen) has supported the opinion that the collagenous fibers are a decisive factor of T2 relaxation time in articular cartilage. Xia et al [37] propose two pools of water molecules associated with two different fractions of collagenous fibers with perpendicular orientation to explain the not completely isotropic behavior of T2 relaxation in intermediate layers of cartilage. From the studies [40,41,48] that have imaged cartilage samples in various orientations to the main magnetic field and that have conducted direct morphologic correlation (scanning or transmission electron microscopy, polarized light microscopy), however, it is not clear that the orientation of the collagenous fibers is the only mechanism that determines T2 relaxation in cartilage. It has been discussed [39–41] that the collagenous fibers architecture as structure-giving component in cartilage imposes a certain order on the spatial alignment of the proteoglycans and that these, in turn, then influence dipolar interactions of associated water molecules.

T2 relaxation and cartilage morphology

In excised human tibial plateau cartilage samples strictly oriented perpendicular to B_0 Lüsse et al [31] demonstrated a good correlation between inverse water content and T2 relaxation rate ($R^2 = 0.71$) with a mean difference between measured (by drying and weighing) and calculated (based on T2) water content of 1.4% ± 0.3%. Nieminen et al [45] demonstrated quite a good correlation ($r = 0.79$) between the spatial distribution of T2 relaxation time across the depth of bovine knee cartilage samples and the inverse birefringence representing collagenous fiber anisotropy. In contrast, no correlation could be demonstrated between T2 relaxation time and safranin O staining as a measure for proteoglycan content in this study. A comparable correlation was presented for samples from different species and different age groups [49]. A decrease of T2 relaxation time was observed (age 4 weeks to 6 months) during the process of skeletal maturation in rats [50].

Treatment with collagenase significantly increased (by up to 200%) T2 relaxation time in the superficial 10% of cartilage in contrast to no significant effect of chondroitinase ABC [51] suggesting sensitivity of T2 relaxation time for the collagenous fibers. This increase of T2 was attributed to a local accumulation of free fluid and an altered orientation pattern of the collagenous fibers. Both types of digestion reduced compressive stiffness of the cartilage. Compression (40%–50% strain) of both normal and trypsin-treated bovine calf patellar cartilage led to a reduction of T2 of approximately 40% [52]. Although trypsin treatment did not alter T2 values in bovine patellar cartilage [53], an increase of T2 was described subsequent to hyaluronidase treatment in rat patellar cartilage [54]. Conversely, in nanomelic, aggrecan-deficient immature chicken cartilage, a significantly lower T2 was observed as compared with normal immature chicken cartilage [55]. T2 relaxation time in the superficial 10% ($r = -0.74$) of intact cartilage samples correlated better with compressive stiffness than overall ($r = -2.1$) T2 [56]. Wayne et al [57] point out that alterations caused by enzymatic treatment must be interpreted cautiously because selective degradation of one specific component of the cartilage matrix is unlikely to occur. In their study average T2 showed a correlation of $R^2 = 0.51$ ($P < .001$) to compressive stiffness and a moderate ($R^2 = 0.44$; $P < .00$) correlation between T2 and average proteoglycan content was observed. Menezes et al [58] underline (and by doing so give a summary on the current experiences) that T2 relaxation time is not specific to one single cartilage component but that it is determined by hydration related to composition and by structure. Kurkijärvi et al [59] emphasize considerable topographic T2 variations in human knee cartilage samples.

T2 relaxation time (as other MR imaging parameters), however, has been shown to be sensitive to biologically relevant changes occurring in cartilage degeneration. Parallel to a reduced compressive stiffness in spontaneously degenerated bovine patellar cartilage, superficial (upper 80 μm) T2 increased from 46 ± 15 milliseconds (intact) over 61 ± 52 milliseconds (moderate OA) to up to 128 ± 109 milliseconds (advanced OA). In contrast, average T2 relaxation time determined over the total cross-section of the cartilage showed an only moderate increase from 46 ± 6 milliseconds (intact) to 59 ± 34 milliseconds (advanced OA) [60]. Similarly, areas of superficial fibrillation showed increased T2 values in cartilage from OA patients [61]. Average rat femorotibial cartilage T2 relaxation time in mechanically induced OA showed a significant increase (by approximately 40%–70%) as compared with the unoperated side [62]. In human cartilage specimen a heterogeneous pattern was observed with a higher T2 in histologically moderate OA than in advanced OA in some samples [63]. Successful spontaneous recovery or repair from artificial focal cartilage defects in rats showed a reduction of initially homogeneously increased T2 of repair tissue to levels of adjacent cartilage over a 60-day period together with reappearance of a zonal differentiation of T2 relaxation time values [64].

Experience from clinical imaging

Current experiences with in vivo measurement of T2 relaxation times are based on data of the knee cartilage. The most ample experience is published by the group [18,19,22–27] using a 3-T scanner and an 8 to 11 echo CPMG approach. To reduce stimulated echoes, the first echo is excluded from evaluation [22]. Spatial resolution is between 0.3 and 0.6 mm in plane and section thickness mostly is chosen at 3 mm [2–4] in one to eight either axial or sagittal sections. In addition to T2 relaxation time maps the authors calculate the T2 relaxation time variations from the cartilage subchondral bone interface to the cartilage surface along radial trajectories. In these T2 profiles, the respective position (depth) in the cartilage is given as the normalized distance from the cartilage subchondral bone interface.

In patella, tibia, and weight-bearing regions of the femur the range of T2 relaxation times is in average between 30 and 67 milliseconds [18,22,24,27] or between 10 and 60 milliseconds [65] in the patellae of healthy volunteers under 45 years. Highest values are found in the upper 20% of the cartilage; lowest values in the lower 30% of the cartilage. Local increase of T2 relaxation time toward the subchondral bone interface in part is attributed to partial volume averaging and chemical shift effects in this area [27]. Variation of T2 relaxation times across the depth of cartilage is more pronounced in the patellae than in femur or tibia [24,27]. A comparable range of T2 relaxation time values in the deep (30–45 milliseconds) and superficial (41–65 milliseconds) portions of cartilage has been described by van Breuseghem et al [66] using a combined inversion recovery spin echo technique to simultaneously obtain T1 and T2 relaxation time data (calculated from two echoes at 7.4 and 70 milliseconds) in healthy volunteers (23–45 years). Comparison of this technique with spectroscopic data from phantom experiments yielded an underestimation of T2 relaxation time by approximately 10% for a range of T2 relaxation time between 9 and 91 milliseconds [66]. Gender in one study did not seem to affect average T2 relaxation time and variation of T2 relaxation time across the cartilage in the knee [24]. In a preliminary study 30 minutes of running led to a decrease of T2 relaxation time (by approximately 6 milliseconds) in the superficial portions of weight-bearing regions of the femoral condyles, whereas in the tibiae no changes could be determined [26]. Data on reproducibility of average patellar cartilage T2 relaxation time are available from two consecutive measurements in 3 out of 20 healthy young adult volunteers [65] before knee bends. The root mean square average coefficient of variation was 1.7%. Average patellar cartilage T2 relaxation time 45 minutes after 60 knee bends was increased by 2.6% ± 1% as compared with 8 minutes after the knee bends. It was attributed to a reuptake of water during recovery of cartilage from loading.

In children (5–17 years) a range of T2 relaxation times comparable with adults has been observed [19] at 1.5 T. In this study, the range of T2 relaxation times of the patella was found to be between 37 and 54 milliseconds and 45 and 62 milliseconds in two groups differing by 6 years of average age where the younger group was imaged in the axial and the older group in the sagittal plane. Physeal cartilage was found to show higher T2 relaxation times (70–80 milliseconds) [19]. Compared with healthy children there was an average increase (by 5 milliseconds) of T2 relaxation time in femoral cartilage of children suffering from juvenile rheumatoid arthritis [67].

Higher age in adults was shown to correlate with increased T2 relaxation times in cartilage [25]. Compared with healthy volunteers between 18 and 30 years, there was a statistically significant increase of T2 relaxation time in the upper 40% of cartilage (76 ± 11 milliseconds) in asymptomatic volunteers aged from 46 to 65 years. In asymptomatic volunteers from 66 to 86 years such an increase was found throughout the whole cartilage (maximum T2 relaxation times: 80 ± 7 milliseconds). It is discussed by the authors that there may be overlap in T2 relaxation times between effects of aging and early preclinical degenerative changes affecting T2 relaxation time in an otherwise healthy elder population.

Symptomatic patients who have findings indicative of chondromalacia patellae showed a more heterogeneous T2 relaxation time distribution in patellar cartilage and foci of increased T2 relaxation time of up to 100 milliseconds [18,22]. Comparison of seven healthy volunteers with patients suffering from radiographically determined moderate (N = 20) and severe (N = 28) OA in the femorotibial compartment showed a significant increase of average T2 relaxation time for the medial femorotibial and the lateral femoral compartment in OA [68] using a two-echo (10 milliseconds, 45 milliseconds) approach. The range of T2 relaxation times was 34 to 41 milliseconds in OA as opposed to 32 to 35 milliseconds in healthy volunteers. No statistical difference in T2 between moderate and severe OA could be detected. Correlation (Spearman) was assessed between T2 relaxation time and WOMAC score and cartilage volume and thickness yielding values in the medial femorotibial compartment between 0.29 and 0.40 ($P < .05$) for pain and function and values between -0.30 and -0.53 ($P < .05$) for volume and thickness. A positive linear correlation has been described in 16 OA patients (eight male, eight female) between femorotibial cartilage T2 relaxation time and the collagen neoepitope C2C antigen blood levels as a serum marker in rheumatoid arthritis [69].

Diffusion-weighted imaging: technical aspects

Tissue analysis using diffusion-weighted imaging is based on the assumption that magnitude and direction of the local diffusivity in tissue are influenced by the macromolecular environment of the diffusing bulk water. Measuring the spatial restriction of diffusivity (in contrast to unrestricted diffusion in free water) according to the tissue's ultrastructure gives information on the tissue's structural properties.

A very successful and commonly applied technique to measure diffusion is the Stejskal Tanner pulsed gradient spin echo method introducing a pair of additional diffusion-sensitizing gradients before and after the refocusing radiofrequency pulse [70]. Signal (S) attenuation is related to additional spin dephasing because of diffusional movement of the water protons that is not refocused before read out. The amount of this signal attenuation is proportional to the amount of diffusion in the tissue, quantified by the apparent diffusion coefficient (ADC), and to the diffusion weighing (b) of the sequence: $S(b) = S_0 \exp(-b \times ADC)$. The degree of diffusion weighing depends on the strength (ie, amplitude and duration) of the diffusion gradients and on the time interval (Δ, the so-called "diffusion time") between these gradients, allowing for diffusion-related dephasing to

occur. Pixel-wise calculation of the ADC results in an ADC map reflecting local diffusional properties throughout the cross-section of cartilage.

With respect to the directional information of diffusion one has to bear in mind that in conventional diffusion-weighted imaging the diffusion sensitizing gradients are applied in but one direction and that

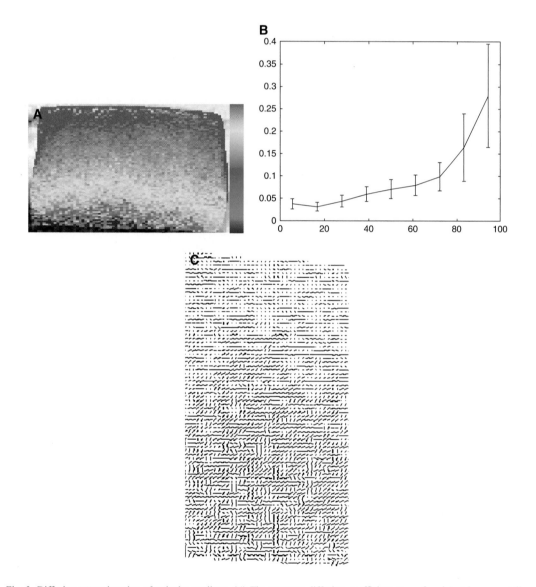

Fig. 3. Diffusion tensor imaging of articular cartilage. (*A*) The apparent diffusion coefficient map of an intact human patellar cartilage sample shows a gradual decrease from the surface down to the tide mark. There is no marked gradient in the tangential plane. (*B*) The profile of fractional anisotropy across the cartilage (normalized depth from the surface) indicates almost no anisotropy at a depth of approximately 15% to 20% with only minimal increase toward the surface. In the depth of the cartilage, however, there is a marked increase in fractional anisotropy. (*C*) The plot of the largest eigenvector shows the distribution of local predominant direction of diffusion. It exhibits a layer with clear predominance of tangential orientation in the upper portion and a more radial alignment in the lower portion of the cartilage, similar to the alignment of the collagenous fibers.

consequently only the component of the total diffusional movement in this direction can be registered. This restriction is overcome with diffusion tensor imaging (DTI) by applying several diffusion-sensitizing gradient pairs in different noncoplanar directions. If six or more gradient directions are available enough information can be obtained completely to evaluate the various directional components of the diffusion pathway [71,72]. These spatially oriented informations can be obtained pixel-wise and collected in a symmetric 3 × 3 data matrix, the diffusion tensor. Diagonalization of this tensor allows calculating the three orthogonal "eigenvectors" and their absolute values, the "eigenvalues," of the tensor. They represent the three main axes (directions) of diffusion corresponding to the three main axes of anisotropy in the tissue. In contrast to T2 relaxation times the eigenvectors are able to provide directional information in various orientations in addition to nondirectional anisotropy without manipulating the probe relative to B_0. The largest eigenvector represents the predominant local (voxel) direction of diffusion (related structural anisotropy) in the probe. From the diffusion tensor the ADC can be calculated as a scalar quantity defined as the mean of the three eigenvalues. As a measure of anisotropy, such parameters as the fractional anisotropy can be calculated as the amount of anisotropic diffusion within the tensor normalized to the modulus of the tensor with values in the interval [0, 1]. Going beyond conventional diffusion-weighted MR imaging, DTI allows one to determine the degree of diffusional anisotropy and the main directions of local diffusion in a tissue (Fig. 3) [71].

Technically, application of diffusion-weighted imaging to cartilage requires low sensitivity-to-susceptibility differences, which is provided by pulsed gradient spin echo sequences. They require acquisition times of several minutes for each diffusion-sensitizing gradient direction to obtain sufficient signal-to-noise ratios (by applying not too short TRs), however, and are very sensitive to motion. An approach to overcome this motion sensitivity is to acquire additional echoes (the navigator echoes), which are used to adjust for inconsistent phase information [73–75]. Recently, first in vivo measurements at 1.5 T of patellar cartilage ADC (one direction; $0.5 \times 0.7 \times 3$ mm^3 resolution, $256 \times 192 \times 16$ matrix) using a three-dimensional steady state sequence [76] with a nonlinear three-dimensional navigator technique [77] have been presented. Brihuega-Moreno et al [78] have proposed a theoretical approach to optimize the b-value scheme with regard to acquisition time and precision of ADC calculation in cartilage.

Diffusion-weighted imaging and cartilage morphology

In view of analysis of a tissue's structural anisotropy, the diffusion time (Δ) may play an important role. Burstein et al [79] described a 40% reduction of diffusivity in intact, trypsin-treated, and compressed cartilage samples because of increased (from 25–2000 milliseconds) diffusion times. Whereas diffusivity in cartilage was restricted to approximately 60% of free water diffusivity at short diffusion times, it was restricted down to only approximately 40% of free water diffusivity at long diffusion times. This indicates that diffusion restricting structural properties of the cartilage matrix can be emphasized and made visible more clearly when longer diffusion times are used. Knauss et al [80] suggest that short and long diffusion times may primarily reflect water content and properties of the collagenous matrix of cartilage, respectively.

The ADC of cartilage has been reported to increase from between 0.68 and 0.75×10^{-3} mm^2/s close to the tide mark to between 1.20 and 1.45×10^{-3} mm^2/s close to the cartilage surface in excised plugs of calf, canine humeral head, and human patellar and femoral condyle cartilage [79,81,82]. The decrease of diffusivity was most pronounced in the lower 30% to 50% of the cartilage samples [82]. In vivo diffusion measurements are expected to yield higher ADC values because they are conducted at a higher temperature (37°C body temperature as opposed to 20°–25°C room temperature) than ex vivo experiments [83]. Budinsky et al [83] demonstrated a linear relationship between (not spatially resolved) cartilage water content of 60% to 80% and cartilage diffusivity as normalized to free water diffusivity. According to Burstein et al [79], matrix charge did not affect diffusivity, whereas compression (35% strain) led to a 19% decrease of cartilage diffusivity.

Trypsin digestion [79,84,85], hyaluronidase, and collagenase digestion [84] led to an increase (10%–30%) of bulk ADC in contrast to retinoic acid digestion [84]. Xia et al [84] reported on concomitant proteoglycan loss measured by the dimethylmethylene blue assay [86] in cartilage exclusively treated with collagenase. According to Toffanin et al [87], however, diffusivity in cartilage was reduced by applying proteoglycan-extracting agents. One theory to explain the observed increase of ADC subsequent to enzymatic treatment is that removal of macromolecules from the cartilage matrix may create pores at the molecular level in the tissue facilitating diffusional movement [84,85].

Qualitatively, in a cadaveric specimen, ADC was increased at 1.5 T in an area of softening as compared with adjacent normal patellar cartilage [88]. In an experimental scanner (7 T), ADC was found to be elevated by approximately 10% throughout the whole depth of cartilage in OA as compared with normal canine humeral cartilage samples [84]. Recently, Mlynarik et al [81] showed an increase of ADC by 30% to 40% in regions of short T1 relaxation times and low proteoglycan staining from OA cartilage as compared with adjacent normal cartilage. This relationship could only be observed in two thirds of their samples, however, whereas in the remaining third no differences in ADC could be observed in the areas of proteoglycan loss, consistent with the assumption that altered ADC values reflect structural degradation of the cartilage matrix.

Going beyond the assessment of the spatial distribution of (nondirectional) ADC in cartilage, Wentorf et al [89] applied diffusion-sensitizing gradients parallel and perpendicular to the cartilage surface in human femoral and bovine patellar cartilage samples, indicating variations of ADC in both directions with increasing distance from the cartilage surface. Variation of ADC was between 1.1×10^{-3} mm^2/s and 0.8×10^{-3} mm^2/s. In an initial study applying DTI to cartilage specimen Filidoro et al [90] demonstrated lowest fractional anisotropy (0.04) at a depth of 20% from the surface clearly increasing (to 0.27) close to the tide mark. Mean diffusivity decreased from the surface ($1.28 \times 10^{-3} \pm 0.14$ mm^2/s) to the tide mark ($0.74 \times 10^{-3} \pm 0.19$ mm^2/s). The alignment of the largest eigenvector showed a high similarity to the zonal alignment of the collagenous fibers as reported from scanning electron microscopy data. Given its high degree of internal structural anisotropy, articular cartilage seems a very interesting and potentially rewarding object to be studied by DTI especially in view of the spatial organization of its collagenous fiber network.

Summary

A comprehensive analysis of articular cartilage's matrix is challenging because of the complex structural relationships of its constituents and because of limited availability of comprehensive and specific biochemical and morphologic analytic methods. In view of recent therapeutic approaches to cartilage damage in OA, however, it is potentially rewarding to develop and further refine noninvasive quantitative tools for specific diagnosis and follow-up studies. In this connection, the most promising approach seems to combine various techniques to obtain the best evaluation possible. Candidates are T2 relaxation time assessment and diffusion-weighted imaging.

Meanwhile, there is considerable experimental and some clinical experience with T2 relaxation time measurements. The T2 values are related to structural properties of the collagenous fiber network and seem to be sensitive to alterations of its integrity and to alterations of its water content, both associated with reduced mechanical stiffness. Their sensitivity to proteoglycan depletion is not yet clear. There is topographic variation of T2 across joint surfaces. T2 seems to decrease from birth to late childhood with developing skeletal maturity and to increase compared with early adulthood in elderly people. T2 relaxation time is significantly increased in areas of degeneration but current experience does not allow a clear-cut differentiation between aging and OA. Acquisition protocols have to be taken into account when comparing values from different centers. In addition to global values it is important to resolve the spatial distribution of T2 over the cartilage in view of detection and follow-up of developing lesions in cartilage.

Motivation for diffusion-weighted imaging and diffusion tensor imaging as comparably new techniques for cartilage imaging is to obtain directly additional architectural and directional information about the cartilage matrix. An important parameter that might help to differentiate between alterations primarily caused by water content versus alterations primarily related to structural matrix effects is the diffusion time Δ. First results suggest that ADC is sensitive to early degenerative damage of cartilage. DTI eigenvectors have additional potential to resolve three-dimensional orientation cartilage matrix properties.

Acknowledgments

Special thanks are given to Dipl. Phys. T. Mendlik and Dr. rer. nat. O. Dietrich for their support in finalizing the manuscript and for the many fruitful discussions about the topic of cartilage imaging.

References

[1] Kempson GE The mechanical properties of articular cartilage. In: Sokoloff L, editor. The joints and synovial fluid, vol 2. 1980. p. 177–237.
[2] Radin EL, Burr DB, Caterson B, et al. Mechanical

determinants of osteoarthrosis. Semin Arthritis Rheum 1991;31(3 Suppl 2):12–21.

[3] Gabriel SE, Crowson CS, Campion ME, et al. Indirect and nonmedical costs among people with rheumatoid arthritis and osteoarthritis compared with nonarthritic controls. J Rheumatol 1997;24:43–8.

[4] MacLean C, Knight K, Paulus H, et al. Costs attributable to osteoarthritis. J Rheumatol 1998;25: 2213–8.

[5] Yelin E, Callaghan LF. The economic cost and social and psychological impact of musculoskeletal conditions. Arthritis Rheum 1995;38:1351–62.

[6] Hunziker EB. Articular cartilage repair: basic science and clinical progress. A review of the current status and prospects. Osteoarthritis Cartilage 2002;10:432–63.

[7] Akizuki S, Mow VC, Muller F, et al. Tensile properties of human knee joint cartilage. II. Correlations between weight bearing and tissue pathology and the kinetics of swelling. J Orthop Res 1987;5:173–86.

[8] Akizuki S, Mow VC, Muller F, et al. Tensile properties of human knee joint cartilage: I. Influence of ionic conditions, weight bearing, and fibrillation on the tensile modulus. J Orthop Res 1986;4:379–92.

[9] Glaser C, Putz R. Functional anatomy of articular cartilage under compressive loading: quantitative aspects of global, local and zonal reactions of the collagenous network with respect to the surface integrity. Osteoarthritis Cartilage 2002;10:83–99.

[10] Buckwalter JA, Mankin HJ. Articular cartilage: degeneration and osteoarthritis, repair, regeneration, and transplantation. Instr Course Lect 1998;47:477–504.

[11] Clarke JC. Articular cartilage: a review and scanning electron microscope study. J Bone Joint Surg Br 1971; 53:732–50.

[12] Clark JM. The organisation of collagen fibrils in the superficial zones of articular cartilage. J Anat 1990; 171:117–30.

[13] Putz R, Fischer H. Altersunterschiede der anordnung der kollagenen fasern im gelenkknorpel. Osteologie Aktuell 1993;7:42–4.

[14] Meachim G, Denham D, Emery IH, et al. Collagen alignments and artificial splits at the surface of human articular cartilage. J Anat 1974;118:101–18.

[15] O'Connor P, Orford CR, Gardner DL. Differential response to compressive loads of zones of canine hyaline articular cartilage: micromechanical, light and electron microscopic studies. Ann Rheum Dis 1988; 47:414–20.

[16] Grushko G, Schneiderman R, Maroudas A. Some biochemical and biophysical parameters for the study of the pathogenesis of osteoarthritis: a comparison between the processes of ageing and degeneration in human hip cartilage. Connect Tissue Res 1989;19: 149–76.

[17] Venn M, Maroudas A. Chemical composition and swelling of normal and osteoarthrotic femoral head cartilage. I. Chemical composition. Ann Rheum Dis 1977;36:121–9.

[18] Dardzinski BJ, Mosher TJ, Li S, et al. Spatial variation of T2 in human articular cartilage. Radiology 1997; 205:546–50.

[19] Dardzinski BJ, Laor T, Schmithorst VJ, et al. Mapping T2 relaxation time in the pediatric knee: feasibility with a clinical 1.5-T MR imaging system. Radiology 2002;225:233–9.

[20] Maier CF, Tan SG, Hariharan H, et al. T2 quantitation of articular cartilage at 1.5 T. J Magn Reson Imaging 2003;17:358–64.

[21] Mendlik T, Faber SC, Weber J, et al. T2 quantitation of human articular cartilage in a clinical setting at 1.5 T: implementation and testing of four multiecho pulse sequence designs for validity. Invest Radiol 2004;39: 288–99.

[22] Mosher TJ, Dardzinski BJ, Smith MB. Human articular cartilage: influence of aging and early symptomatic degeneration on the spatial variation of T2–preliminary findings at 3 T. Radiology 2000;214:259–66.

[23] Mosher TJ, Smith H, Dardzinski BJ, et al. MR imaging and T2 mapping of femoral cartilage: in vivo determination of the magic angle effect. AJR Am J Roentgenol 2001;177:665–9.

[24] Mosher TJ, Collins CM, Smith HE, et al. Effect of gender on in vivo cartilage magnetic resonance imaging T2 mapping. J Magn Reson Imaging 2004;19: 323–8.

[25] Mosher TJ, Liu Y, Yang QX, et al. Age dependency of cartilage magnetic resonance imaging T2 relaxation times in asymptomatic women. Arthritis Rheum 2004; 50:2820–8.

[26] Mosher TJ, Smith HE, Collins C, et al. Change in knee cartilage T2 at MR imaging after running: a feasibility study. Radiology 2005;234:245–9.

[27] Smith HE, Mosher TJ, Dardzinski BJ, et al. Spatial variation in cartilage T2 of the knee. J Magn Reson Imaging 2001;14:50–5.

[28] Duewell SH, Ceckler TL, Ong K, et al. Musculoskeletal MR imaging at 4 T and at 1.5 T: comparison of relaxation times and image contrast. Radiology 1995; 196:551–5.

[29] Mlynarik V, Szomolanyi P, Toffanin R, et al. Transverse relaxation mechanisms in articular cartilage. J Magn Reson 2004;169:300–7.

[30] Gold GE, Han E, Stainsby J, et al. Musculoskeletal MRI at 3.0 T: relaxation times and image contrast. AJR Am J Roentgenol 2004;183:343–51.

[31] Lüsse S, Claassen H, Gehrke T, et al. Evaluation of water content by spatially resolved transverse relaxation times of human articular cartilage. Magn Reson Imaging 2000;18:423–30.

[32] Poon CS, Henkelman RM. Practical T2 quantitation for clinical applications. J Magn Reson Imaging 1992; 2:541–53.

[33] Does MD, Snyder RE. Multiecho imaging with suboptimal spoiler gradients. J Magn Reson 1998;131: 25–31.

[34] Henkelman RM, Stanisz GJ, Kim JK, et al. Anisotropy of NMR properties of tissues. Magn Reson Med 1994; 32:592–601.

[35] Mlynarik V, Degrassi A, Toffanin R, et al. Investigation of laminar appearance of articular cartilage by means of magnetic resonance microscopy. Magn Reson Imaging 1996;14:435–42.

[36] Lüsse S, Knauss R, Werner A, et al. Action of compression and cations on the proton and deuterium relaxation in cartilage. Magn Reson Med 1995;33: 483–9.

[37] Xia Y, Moody JB, Alhadlaq H. Orientational dependence of T2 relaxation in articular cartilage: a microscopic MRI (microMRI) study. Magn Reson Med 2002;48:460–9.

[38] Kneeland JB. Articular cartilage and the magic angle effect. AJR Am J Roentgenol 2001;177:671–2.

[39] Xia Y. Relaxation anisotropy in cartilage by NMR microscopy (muMRI) at 14-microm resolution. Magn Reson Med 1998;39:941–9.

[40] Goodwin DW, Wadghiri YZ, Dunn JF. Micro-imaging of articular cartilage: T2, proton density, and the magic angle effect. Acad Radiol 1998;5:790–8.

[41] Goodwin DW, Zhu H, Dunn JF. In vitro MR imaging of hyaline cartilage: correlation with scanning electron microscopy. AJR Am J Roentgenol 2000;174:405–9.

[42] Erickson SJ, Prost RW, Timins ME. The "magic angle" effect: background physics and clinical relevance. Radiology 1993;188:23–5.

[43] Hayes CW, Parellada JA. The magic angle effect in musculoskeletal MR imaging. Top Magn Reson Imaging 1996;8:51–6.

[44] Gründer W, Wagner M, Werner A. MR-microscopic visualization of anisotropic internal cartilage structures using the magic angle technique. Magn Reson Med 1998;39:376–82.

[45] Nieminen MT, Rieppo J, Toyras J, et al. T2 relaxation reveals spatial collagen architecture in articular cartilage: a comparative quantitative MRI and polarized light microscopic study. Magn Reson Med 2001;46: 487–93.

[46] Mlynarik V. Magic angle effect in articular cartilage. AJR Am J Roentgenol 2002;178:1287.

[47] Goodwin DW, Dunn JF. MR imaging and T2 mapping of femoral cartilage. AJR Am J Roentgenol 2002;178: 1568–9.

[48] Xia Y, Moody JB, Alhadlaq H, et al. Imaging the physical and morphological properties of a multi-zone young articular cartilage at microscopic resolution. J Magn Reson Imaging 2003;17:365–74.

[49] Nissi MJ, Rieppo J, Töyräs J, et al. T2 relaxation reveals differences in spatial collagen network anisotropy in human, bovine and porcine articular cartilage [abstract]. In: Proceedings of the International Society of Magnetic Resonance Medicine 2003;11. ISMRM, Toronto, Canada, July 10–16, 2003. Available at: http://www.ismrm.org/03/.

[50] Watrin A, Ruaud JP, Olivier PT, et al. T2 mapping of rat patellar cartilage. Radiology 2001;219:395–402.

[51] Nieminen MT, Toyras J, Rieppo J, et al. Quantitative MR microscopy of enzymatically degraded articular cartilage. Magn Reson Med 2000;43:676–81.

[52] Kaufman JH, Regatte RR, Bolinger L, et al. A novel approach to observing articular cartilage deformation in vitro via magnetic resonance imaging. J Magn Reson Imaging 1999;9:653–62.

[53] Regatte RR, Akella SV, Borthakur A, et al. Proteoglycan depletion-induced changes in transverse relaxation maps of cartilage: comparison of T2 and T1rho. Acad Radiol 2002;9:1388–94.

[54] Watrin-Pinzano A, Ruaud JP, Olivier P, et al. Effect of proteoglycan depletion on T2 mapping in rat patellar cartilage. Radiology 2005;234:162–70.

[55] Mosher TJ, Chen Q, Smith MB. 1H magnetic resonance spectroscopy of nanomelic chicken cartilage: effect of aggrecan depletion on cartilage T2. Osteoarthritis Cartilage 2003;11:709–15.

[56] Nieminen MT, Toyras J, Laasanen MS, et al. Prediction of biomechanical properties of articular cartilage with quantitative magnetic resonance imaging. J Biomech 2004;37:321–8.

[57] Wayne JS, Kraft KA, Shields KJ, et al. MR imaging of normal and matrix-depleted cartilage: correlation with biomechanical function and biochemical composition. Radiology 2003;228:493–9.

[58] Menezes NM, Gray ML, Hartke JR, et al. T2 and T1rho MRI in articular cartilage systems. Magn Reson Med 2004;51:503–9.

[59] Kurkijarvi JE, Nissi MJ, Kiviranta I, et al. Delayed gadolinium-enhanced MRI of cartilage (dGEMRIC) and T2 characteristics of human knee articular cartilage: topographical variation and relationships to mechanical properties. Magn Reson Med 2004;52:41–6.

[60] Nissi MJ, Toyras J, Laasanen MS, et al. Proteoglycan and collagen sensitive MRI evaluation of normal and degenerated articular cartilage. J Orthop Res 2004;22: 557–64.

[61] Mlynarik V, Trattnig S, Huber M, et al. The role of relaxation times in monitoring proteoglycan depletion in articular cartilage. J Magn Reson Imaging 1999;10: 497–502.

[62] Spandonis Y, Heese FP, Hall LD. High resolution MRI relaxation measurements of water in the articular cartilage of the meniscectomized rat knee at 4.7 T. Magn Reson Imaging 2004;22:943–51.

[63] David-Vaudey E, Ghosh S, Ries M, et al. T2 relaxation time measurements in osteoarthritis. Magn Reson Imaging 2004;22:673–82.

[64] Watrin-Pinzano A, Ruaud JP, Cheli Y, et al. T2 mapping: an efficient MR quantitative technique to evaluate spontaneous cartilage repair in rat patella. Osteoarthritis Cartilage 2004;12:191–200.

[65] Liess C, Lusse S, Karger N, et al. Detection of changes in cartilage water content using MRI T2-mapping in vivo. Osteoarthritis Cartilage 2002;10:907–13.

[66] Van Breuseghem I, Bosmans HT, Elst LV, et al. T2 mapping of human femorotibial cartilage with turbo mixed MR imaging at 1.5 T: feasibility. Radiology 2004;233:609–14.

[67] Kight AC, Dardzinski BJ, Laor T, et al. Magnetic resonance imaging evaluation of the effects of juvenile

rheumatoid arthritis on distal femoral weight-bearing cartilage. Arthritis Rheum 2004;50:901 – 5.

[68] Dunn TC, Lu Y, Jin H, et al. T2 relaxation time of cartilage at MR imaging: comparison with severity of knee osteoarthritis. Radiology 2004;232:592 – 8.

[69] King KB, Lindsey CT, Dunn TC, et al. A study of the relationship between molecular biomarkers of joint degeneration and the magnetic resonance-measured characteristics of cartilage in 16 symptomatic knees. Magn Reson Imaging 2004;22:1117 – 23.

[70] Stejskal EO, Tanner JE. Spin diffusion measurements: spin echoes in the presence of a time-dependent field gradient. J Chem Phys 1965;42:288 – 92.

[71] Le Bihan D, Mangin JF, Poupon C, et al. Diffusion tensor imaging: concepts and applications. J Magn Reson Imaging 2001;13:534 – 46.

[72] Basser PJ, Pierpaoli C. A simplified method to measure the diffusion tensor from seven MR images. Magn Reson Med 1998;39:928 – 34.

[73] Ordidge RJ, Helpern JA, Qing ZX, et al. Correction of motional artifacts in diffusion-weighted MR images using navigator echoes. Magn Reson Imaging 1994;12: 455 – 60.

[74] Anderson AW, Gore JC. Analysis and correction of motion artifacts in diffusion weighted imaging. Magn Reson Med 1994;32:379 – 87.

[75] Dietrich O, Heiland S, Benner T, et al. Reducing motion artefacts in diffusion-weighted MRI of the brain: efficacy of navigator echo correction and pulse triggering. Neuroradiology 2000;42:85 – 91.

[76] Miller KL, Hargreaves BA, Gold GE, et al. Steady-state diffusion-weighted imaging of in vivo knee cartilage. Magn Reson Med 2004;51:394 – 8.

[77] Miller KL, Pauly JM. Nonlinear phase correction for navigated diffusion imaging. Magn Reson Med 2003; 50:343 – 53.

[78] Brihuega-Moreno O, Heese FP, Hall LD. Optimization of diffusion measurements using Cramer-Rao lower bound theory and its application to articular cartilage. Magn Reson Med 2003;50:1069 – 76.

[79] Burstein D, Gray ML, Hartman AL, et al. Diffusion of small solutes in cartilage as measured by nuclear magnetic resonance (NMR) spectroscopy and imaging. J Orthop Res 1993;11:465 – 78.

[80] Knauss R, Schiller J, Fleischer G, et al. Self-diffusion of water in cartilage and cartilage components as studied by pulsed field gradient NMR. Magn Reson Med 1999;41:285 – 92.

[81] Mlynarik V, Sulzbacher I, Bittsansky M, et al. Investigation of apparent diffusion constant as an indicator of early degenerative disease in articular cartilage. J Magn Reson Imaging 2003;17:440 – 4.

[82] Xia Y, Farquhar T, Burton-Wurster N, et al. Diffusion and relaxation mapping of cartilage-bone plugs and excised disks using microscopic magnetic resonance imaging. Magn Reson Med 1994;31:273 – 82.

[83] Budinsky L, Wachsmuth L, Ghosh S, et al. Navigator echo based motion corrected diffusion imaging of articular cartilage in vivo at 1,5 T [abstract]. In: Proceedings of the International Society of Magnetic Resonance Medicine 2001:9. ISMRM, Glasgow, Scotland, April 21 – 27, 2001. Available at: http:// www.ismrm.org/01/.

[84] Xia Y, Farquhar T, Burton-Wurster N, et al. Self-diffusion monitors degraded cartilage. Arch Biochem Biophys 1995;323:323 – 8.

[85] Berg A, Singer T, Moser E. High-resolution diffusivity imaging at 3.0 T for the detection of degenerative changes: a trypsin-based arthritis model. Invest Radiol 2003;38:460 – 6.

[86] Farndale RW, Sayers CA, Barrett AJ. A direct spectrophotometric microassay for sulfated glycosaminoglycans in cartilage cultures. Connect Tissue Res 1982;9:247 – 8.

[87] Toffanin R, Mlynarik V, Russo S, et al. Proteoglycan depletion and magnetic resonance parameters of articular cartilage. Arch Biochem Biophys 2001;390: 235 – 42.

[88] Frank LR, Wong EC, Luh WM, et al. Articular cartilage in the knee: mapping of the physiologic parameters at MR imaging with a local gradient coil – preliminary results. Radiology 1999;210:241 – 6.

[89] Wentorf F, Chen W, Zhang X. initial findings of diffusion anisotropy in articular cartilage [abstract]. In: Proceedings of the International Society of Magnetic Resonance Medicine 2003:11. ISMRM, Toronto, Canada, July 10 – 16, 2003. Available at: http://www.ismrm.org/03/.

[90] Filidoro L, Dietrich O, Weber J, et al. High resolution diffusion tensor imaging of human patellar cartilage: feasibility and preliminary findings. Magn Reson Med, in press.

ELSEVIER
SAUNDERS

Radiol Clin N Am 43 (2005) 655–672

RADIOLOGIC
CLINICS
of North America

MR Imaging of Epiphyseal Lesions of the Knee: Current Concepts, Challenges, and Controversies

Frédéric E. Lecouvet, MD, PhD*, Jacques Malghem, MD,
Baudouin E. Maldague, MD, Bruno C. Vande Berg, MD, PhD

*Section of Musculoskeletal Radiology, Department of Radiology, Saint Luc University Hospital, Université de Louvain,
Hippocrate Avenue 10/2942, Brussels B-1200, Belgium*

The use of MR imaging for the work-up of knee disorders has led to the recognition of frequent alterations of the epiphyseal bone and marrow. Whereas most of these alterations result from cartilage lesions or represent posttraumatic bone marrow lesions, there are several conditions that primarily involve the subchondral bone plate or the epiphyseal bone marrow. Avascular necrosis (AVN) and a group of lesions presenting the bone marrow edema (BME) pattern are the two main categories of conditions that may involve the knee, although they most commonly involve the hip. MR imaging enables distinction between these two main categories, and further categorization of the latter group into spontaneous osteonecrosis of the knee (SONK), and self-resolving conditions with indistinct and probably overlapping borders, presenting the observation of transient BME as a common MR imaging feature.

This article focuses on all these nontraumatic and nondegenerative entities involving mature knee epiphyses. The proposed hypotheses on causes and pathogenesis of AVN, SONK, and self-resolving epiphyseal conditions are discussed. Imaging features that enable early differentiation between irreversible lesions, AVN and SONK, and spontaneously resolutive lesions are emphasized. The most challenging role of the radiologist is to provide a prognosis of the likely lesion outcome early in the course of the disease.

* Corresponding author.
E-mail address: lecouvet@rdgn.ucl.ac.be
(F.E. Lecouvet).

Avascular necrosis

Definition and general features

The term *necrosis* should be regarded differently depending on the target level. At histology, necrosis designates morphologic cellular changes that occur after cell death in a living tissue or organ [1–3]. For the clinician, necrosis of a joint can be defined by the irreversible loss of joint function related to spontaneous collapse of an epiphysis, as demonstrated on conventional radiography. The term *bone infarct*, which typically refers to an organized area of necrosis within a tissue, usually applies to metaphyseal or diaphyseal lesions, although it can also apply to noncollapsed epiphyseal lesions [3].

The knee represents the third most commonly involved joint by AVN, after the hip and the shoulder joints [4]. The frequency of knee AVN is probably underestimated, however, given the lack of large studies based on systematic screening of patients with high risk of AVN. In a limited series of patients with systemic lupus erythematosus, the frequency of knee involvement was similar to that of hip involvement [5].

Knee AVN tends to develop in adults, most commonly in the fourth and fifth decades of life, but age at onset mainly depends on the underlying disease or predisposing factors [4]. In most cases, AVN is secondary and shares the same risk factors as hip AVN. These predisposing conditions include glucocorticoid intake; alcoholism; hyperuricemia; connective tissue disorders (systemic lupus erythematosus);

hemoglobinopathies; and HIV infection [6–10]. In opposition to hip AVN, knee AVN related to knee dislocation or complex knee fracture seems to be rare. Patients with sickle cell anemia tend to have more frequent spinal or hip involvement than knee involvement [11].

The physiopathogeny of knee AVN has not been as widely studied as that of the femoral head. By analogy to accepted theories on the development of femoral head AVN, it is likely that knee AVN develops in response to vascular failure of the bone marrow. Various physiopathologic mechanisms have been proposed to explain this vascular failure, including intravascular thrombosis or embolism, intrinsic vessels disorders, or extravascular compression caused by the development of a relative hyperpressure in a rigid bony compartment [12,13].

Clinical findings

All epiphyses may be involved, especially both femoral condyles, but also metaphyses and diaphyses [14]. Most infarcts, especially of metaphyseal or diaphyseal locations, remain clinically occult, and may even show size regression at follow-up [15]. The clinical presentation of epiphyseal locations also remains long-time silent. The symptoms only occur late and more gradually in comparison with SONK (Table 1) [4]. The occurrence of symptoms in a patient with known bone marrow infarcts could indicate impending subchondral bone fracture at risk for evolution to epiphyseal collapse [16].

Imaging features

Radiographs

Infarcts appear as areas of bone sclerosis mainly located in the metaphysis or epiphysis. At close analysis, bone sclerosis is caused by the presence of a serpiginous osteosclerotic rim that surrounds the infarct. Discrete periosteal reaction may be seen in the metaphyseal area. The two specific radiologic signs for advanced epiphyseal osteonecrosis are an abrupt encroachment of the epiphyseal contour (depression with subsequent loss of epiphyseal sphericity) and the radiolucent crescent sign (area of lucency underlying the subchondral bone plate), which both reflect fracture of the subchondral bone plate (Fig. 1) [8,13,17,18]. These signs are usually observed at the time of onset of symptoms, alone or in association. The crescent sign seems to be much less frequent in knee than in hip AVN, probably

Table 1
Differences between avascular necrosis and spontaneous osteonecrosis of the knee

Characteristic	AVN	SONK
Predisposing factors	Glucocorticoïds, alcohol, connective tissue disorders, AIDS, storage disorders	Age, osteporosis, gender (male > female), obesity
		Meniscal lesion (especially posterior horn of medial meniscus)
Patient age	Any (risk factors)	Elderly
Distribution	Any area (dia-, meta-, epiphysis)	Predilection for medial femoral condyle
	Bilateral, multiple (knee + hip + shoulder)	—
Symptoms	Silent, insidious = infarct	Sudden spontaneous pain
	More acute symptoms = collapse?	Definite moment
Presumed pathogenesis	Vascular origin	Mechanical origin
Radiographic findings	Nothing	Nothing
	Infarct	Variable subchondral bone changes
	Collapse	Collapse
	Radiolucent subchondral area	Radiolucent subchondral area
MR imaging features	Geographic pattern	BME pattern
	Peripheral demarcation rim (low SI on T1, double-line on T2)	Focal abnormalities: subchondral areas of low SI on T2
	±BME[a]	±Fracture lines (low SI)
	±Subchondral fracture cleft (high SI on T2)	±Subchondral fracture cleft (high SI on T2)
	±Subchondral bone plate deformity	±Subchondral bone plate deformity
Histology	Massive bone marrow infarct	Limited subchondral area of necrosis between a fracture line and subchondral bone plate

Abbreviation: SI, signal intensity.
[a] Generally indicative of (impending) epiphyseal collapse.

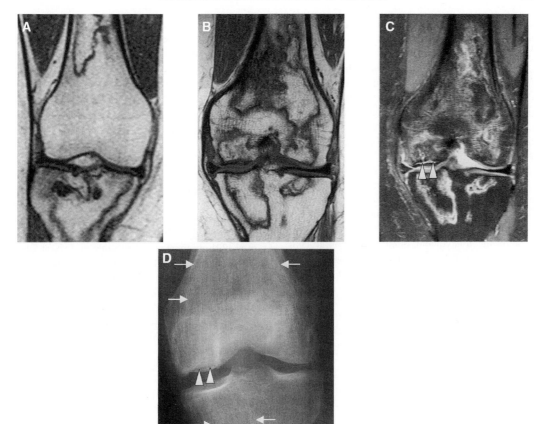

Fig. 1. Multifocal AVN in a 17-year-old patient previously treated with glucocorticoids. (*A*) Coronal spin echo (SE) T1-weighted image of the asymptomatic right knee shows multiple bone marrow infarcts. Coronal SE T1-weighted (*B*) and fat-saturated fast spin echo (FSE) T2-weighted (*C*) images of the symptomatic left knee demonstrate geographic lesions delineated by a characteristic low T1, high T2 serpentine-like peripheral rim. The content of most lesions shows a preserved fatty marrow signal. The lesion involving the medial femoral condyle shows a subchondral fluid-like signal intensity line suggestive of subchondral fracture (*arrowheads*), with flattening of the weight-bearing area of this condyle. (*D*) Anteroposterior (AP) radiograph of the left knee reveals frank depression of the weight-bearing area of the medial femoral condyle and the subchondral radiolucent crescent sign (*arrowheads*) typical for collapsed epiphyseal osteonecrosis. Note discrete peripheral sclerotic rim of femoral and tibial infarcts (*arrows*).

because joint distraction is needed to allow visualization of this sign.

Late-stage AVN with epiphyseal deformity leads to secondary osteoarthritis (OA); it may then become impossible (and without clinical significance) to differentiate advanced AVN with secondary OA from epiphyseal deformity caused by cartilage loss and bone destruction in primary OA.

Bone scintigraphy

Bone scans performed with a bone-seeking agent, such as Tc 99m bisphosphonates, have proved useful to detect AVN. Their sensitivity, however, seems limited at the precollapse stage, before clinical manifestation of the disease, such as in AVN of the femoral head [16]. Characteristic findings of pre-collapse hip AVN (a cold [photopenic] zone and the donut sign defined by the presence of a peripheral increased uptake area in a metaphysis or epiphysis) have not been definitely reported in the knee [19,20]. Once collapse has occurred, diffuse increased activity is generally observed but lacks specificity, for example to differentiate AVN from SONK [4,16].

MR imaging

Several imaging findings have been observed in knee AVN. The most common MR imaging appearance of AVN is that of a geographic lesion of more or less preserved fatty marrow signal (Fig. 1). On T1-weighted images, the lesion is of high signal intensity, surrounded by a low signal intensity rim that is pathognomonic for AVN. At histology, this low signal intensity rim correlates with the reactive interface between viable and nonviable bone marrow. On T2-weighted images, this rim may present a double-line sign, which is characterized by the juxtaposition of low and high signal intensity rims [21,22]. This sign may be absent on fat-saturated images because it could be related to a chemical-shift artifact (Fig. 1) [22–25].

Evolution to epiphyseal necrosis and progressive collapse are associated with changes on MR images: appearance of localized BME around the infarct, which may be responsible for blurring of its limits; signal intensity changes within the weight-bearing area of the infarct; and finally subchondral bone fracture. This subchondral bone fracture appears as a frank depression of the subchondral bone plate with subsequent loss of epiphyseal sphericity, or more rarely as a fluid-like high signal intensity line on T2-weighted images underlying the subchondral bone plate, representing a subchondral fracture cleft (Fig. 1). The signal intensity of the necrotic tissue within the infarct is then equivalent to that of fat or of low signal intensity on both T1- and T2-weighted images, reflecting the presence of either mummified fat or eosinophilic necrosis, respectively [26,27].

Work-up of patients with suspected knee avascular necrosis

Radiographs should be the initial study in the work-up of patients who have clinically suspected AVN. When radiographs explain the symptoms by showing advanced AVN or degenerative disease, further imaging is usually unwarranted. When normal or nonconclusive, radiographs should be followed by an MR imaging study because this technique is sensitive and specific for the depiction of AVN. It also provides valuable information on articular structures (menisci, cartilage, and ligaments).

Staging

An accurate staging system for knee AVN should incorporate as many different features as those used for the staging of hip AVN. To the best of the authors' knowledge, there is no available or widely accepted staging system for knee AVN. In the femoral head, Mitchell et al [28] described four patterns based on the signal intensity of the central portion of the lesion on T1- and T2-weighted images (fat-like signal in class A, hemorrhage-like signal in class B, fluid-like

signal in class C, and fibrous-like signal in class D) [25]. Despite the lack of an available staging system, the report should mention the presence or absence of fracture of the subchondral bone plate, because this occurrence of fracture may represent a critical phase in the natural history of epiphyseal AVN.

Prognostic factors

There are no large studies addressing the value of MR imaging in predicting the risk for epiphyseal collapse in knee AVN, like for hip AVN. There is preliminary evidence that suggests that the overall risk for collapse is lower in AVN lesions of the femoral condyle than in femoral head AVN. Sakai et al [29] observed a collapse rate of 15% at 24- to 50-month follow-up. Despite the lack of large studies, it is likely that the risk of collapse in a femoral condyle is directly related to the size of the lesion and to its location. According to Sakai et al [29], AVN lesions involving more than one third of the condyle on the mid-coronal MR images, or the middle and posterior thirds of this condyle on mid-sagittal MR images, were at higher risk of collapse. In the authors' experience, it is also noteworthy that infarcts involving the tibial plateaus are at lower risk of collapse than lesions involving the femoral condyles.

Treatment

Core decompression has been used by Mont et al [30] in AVN of the femoral condyle and seems to slow the pace of disease progression. When epiphyseal bone plate has fractured, conservative treatment under radiographic follow-up may be proposed until functional joint impairment. In late stages with epiphyseal collapse and secondary OA, knee arthroplasty is often the final treatment option.

Epiphyseal lesions presenting bone marrow edema as common prominent feature

Spontaneous osteonecrosis of the knee

First reported by Ahlbäck et al [31] in 1968, SONK has been identified as a distinct form of epiphyseal osteonecrosis. Indeed, clinical, histologic, and imaging (especially MR imaging) features completely separate SONK from knee AVN (Table 1) [4]. Constructive debate between clinicians, radiologists, and pathologists progressively led to in-depth knowledge of this lesion, to the description of its specific features, and finally to the elaboration of the accepted hypothesis of its mechanical or microtraumatic origin [16,17].

Definition and general features

SONK is observed in the elderly, usually after the sixth decade of life, and more frequently in women than in men (three times more frequently) [4,31–33]. No association is found with systemic or metabolic disorders or therapeutic agents [33]. SONK shows a frank predilection for the medial femoral condyle, involved in more than 90% of cases; rarely, isolated involvement of the lateral femoral condyle and tibial plateaus has been reported [32–40]. SONK is not associated with similar involvement of the contralateral knee or of the hip (Table 1).

Clinical presentation

SONK is usually observed in elderly people and is characterized by a spontaneous knee pain with an acute onset. This pain is so sudden that the patient frequently remembers the exact moment or activity during which the symptoms started. This pain is initially severe, and remains present at rest [31,33, 41,42]. The pace of disease progression depends on its location and size. Small lesions (< 3.5 cm^2) may be well tolerated for several years, whereas extensive lesions with subchondral bone plate fracture result in

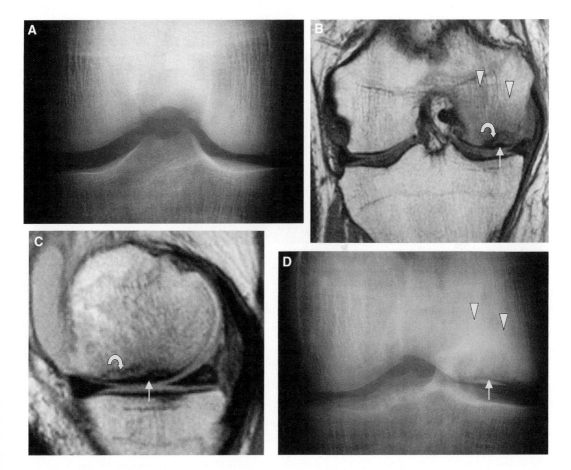

Fig. 2. SONK in a 63-year-old man: early imaging findings. (*A*) AP radiograph of the right knee performed 5 weeks after onset of symptoms shows no evident abnormality. Coronal SE T1-weighted (*B*) and sagittal FSE proton density weighted (*C*) images show a discrete BME pattern (*arrowheads in B*), a subchondral crescent-like area of a low signal intensity on both images (*arrows*) typical for subchondral osteonecrosis, and a deeper low signal intensity line (*curved arrows*) suggestive of a subchondral fracture. (*D*) A 5-month follow-up AP radiograph shows epiphyseal osteonecrosis with a radiolucent crescent sign underlying the epiphyseal surface (*arrow*) surrounded by sclerosis (*arrowheads*).

severe persisting complaints and functional joint impairment [9,43,44].

Imaging features

Radiographs. Radiographic changes are variable according to the time delay between onset of symptoms and radiographic examinations (Figs. 2 and 3) [32,33,37,41]. Initial radiographs are frequently normal, with reported frequencies of radiographic abnormalities at onset of symptoms ranging from 10% to 43% [45].

Several radiographic features have been reported and were used for staging because there seems to be a cascade of events that leads to collapse. These radiographic features are normal radiographs (stage I); slight flattening or subtle focal depression of the subchondral bone plate without joint narrowing (stage II); focal epiphyseal collapse consisting of a frank subchondral bone plate depression or a sub-

chondral radiolucent crescent sign, sometimes surrounded by discrete sclerotic area in the subchondral bone (stage III) or with an evident peripheral sclerotic halo (stage IV); and epiphyseal deformity progressing over time resulting in secondary OA, which may dissimulate the characteristic radiographic features of SONK (stage V) [16,31,33,46].

The prognostic value of radiographic abnormalities has been assessed. The location, size, and width of the lesion are of prognostic value, with a better clinical outcome when the lesion is small [9,41]. Lesions with a width superior to 40% to 50% of the femoral condyle or a surface area superior to 5 cm^2 have an unfavorable outcome with earlier progression to epiphyseal collapse and OA [32,33,46,47].

Bone scintigraphy. Bone scans show increased isotope uptake within the subchondral bone of the involved femoral condyle, but this finding lacks

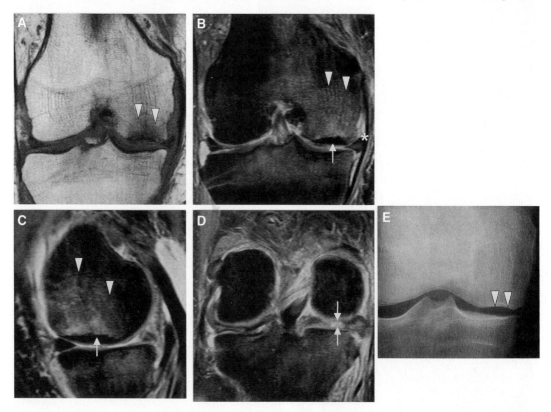

Fig. 3. SONK in a 77-year-old woman: association with meniscal lesion. Coronal SE T1-weighted (*A*) and fat-saturated FSE T2-weighted (*B*) images show ill-delimited BME within the medial femoral condyle (*arrowheads*) and a subchondral crescent-like low signal intensity area on the T2-weighted image (*arrow in B*). Note medial subluxation of the medial meniscus (***). (*C*) Sagittal fat-saturated FSE T2-weighted image shows ill-delimited high signal intensity of BME (*arrowheads*) and an extensive subchondral crescent-like low signal intensity area (*arrow*). (*D*) Coronal fat-saturated FSE T2-weighted image shows deep radial tear of the posterior horn of the medial meniscus (*arrows*). (*E*) A 2-month follow-up AP radiograph shows discrete collapse of the surface of the medial femoral condyle (*arrowheads*).

specificity, being also observed in degenerative disease, AVN and subsequent epiphyseal collapse, insufficiency fractures, and other self-resolving epiphyseal lesions (see later) [16,17,31,32,48].

MR imaging. The most prominent MR imaging characteristics of SONK are the poorly delimited BME pattern and the lack of peripheral low signal intensity rim as seen in AVN (Figs. 2 and 3) [46, 49–52]. The BME pattern shows low signal intensity on T1-weighted images, intermediate to high signal intensity on T2-weighted images, and enhancement after contrast injection. T1-weighted MR images are sufficient in most cases for confident differentiation between SONK and AVN. The BME pattern is not specific for SONK and may be observed in transient epiphyseal conditions (detailed later). More specific features are necessary to distinguish SONK from these conditions, which can be identified on high-resolution T2-weighted MR images in the subchondral bone.

A focal subchondral area of low signal intensity on T2-weighted and proton density weighted images, abutting the subchondral bone plate (and responsible for apparent thickening of this plate), is the most specific MR imaging finding for SONK, found in almost all cases (Figs. 2 and 3) [52]. This area shows no enhancement on postcontrast T1-weighted MR images [45,46]. At macroscopic examination, this area appears as a whitish gray zone underlying the subchondral osseous end plate [17,37]. Microscopic examination of this necrotic area reveals cellular debris and thickened collapsed bone trabeculae, irregularly arranged fracture callus, reactive cartilage, and granulation tissue [17,45,51–53].

Another MR imaging finding suggestive for SONK is the presence of a deformity of the subchondral bone plate (flattening or focal depression) in the weight-bearing area of the involved condyle [52]. A less frequent but also very suggestive finding is the observation of a subchondral fracture cleft of fluid-like high signal intensity on T2-weighted images underlying the epiphyseal end plate, which is interpreted as a sign of subchondral bone fracture, and likely corresponds to the radiolucent crescent sign seen on radiographs (Fig. 4) [52].

Ancillary and nonspecific features frequently observed in SONK include lines of low signal intensity on all sequences running in the subchondral bone at several millimeters from the epiphyseal bone plate (Fig. 2). These lines are reminiscent of the open-ended low signal intensity lines seen in occult or microtraumatic fractures, and are interpreted as insufficiency fractures within the trabecular bone [17,54]. Less frequently, rounded cystic areas with high signal intensity on T2-weighted images are observed in the subchondral area [51].

Pathogenesis

The etiopathogenesis of SONK has been widely discussed in the literature. A mechanical or microtraumatic origin has progressively emerged as the

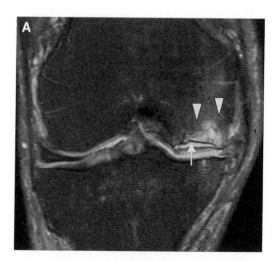
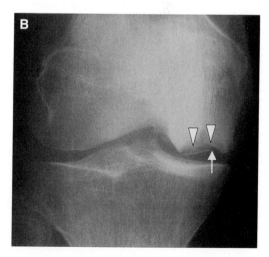

Fig. 4. SONK in a 75-year-old man: subchondral fracture cleft. (*A*) Coronal fat saturated FSE T2-weighted image show ill-delimited BME within the medial femoral condyle (*arrowheads*) and a subchondral fluid-like high signal intensity cleft (*arrow*). (*B*) AP radiograph shows evident epiphyseal osteonecrosis with collapsed epiphyseal surface (*arrowheads*) and subchondral end plate fracture (*arrow*).

most likely underlying mechanism of SONK [4,33, 41,52]. According to this hypothesis, repetitive impact or stress on the epiphyseal bone plate and underlying trabecular network may induce microfractures in the trabecular bone, especially if this bone is weakened by osteopenia [16,41,42,45,46,49, 55,56]. Accumulation of these microfractures with failure of repair mechanisms could induce subchondral bone destruction, which may eventually lead to subchondral bone plate collapse typical for epiphyseal osteonecrosis. Clinical findings, MR imaging, and histology support this hypothesis of the role of a subchondral bone fracture as initiating lesion (Box 1)

Box 1. Features supporting the hypothesis of a mechanical origin of spontaneous osteonecrosis of the knee

Clinical findings

- Older age
- Association with bone weakening (elderly women, osteoporosis, transplant, etc)
- Lack of predisposing factor for AVN
- Frequent obesity

Affected knee

- Predilection for medial femoral condyle (> 90%) and weight-bearing area
- Significant association with meniscal lesions (especially radial tears and root derangements of the posterior horn of the medial meniscus)

MR imaging

- Lack of cardinal features of AVN (demarcation rim)
- Presence of subchondral insufficiency fractures (SIFs) in involved epiphysis
- Association with SIF elsewhere in the knee

Histology

- Limited subchondral area of necrosis
- Underlying SIF

[17,49,57]. The most important clinical factors that promote this hypothesis are the abrupt onset of symptoms, the specific patient population, the absence of systemic risk factor, and the lesion location (Box 1).

The complete difference in the MR imaging appearance of SONK and AVN most likely reflects a difference in pathogenesis. MR images of SONK do not show the serpentine-like peripheral rim typical for AVN, but the BME pattern and more specific subchondral marrow abnormalities representing limited areas of necrosis (focal subchondral low signal intensity areas on T2-weighted images) [51–54]. This distinction of two different MR imaging patterns of epiphyseal osteonecrosis in the knee (AVN and SONK) parallels the distinction made in the hip between typical AVN, presenting a geographic MR imaging pattern, and more exceptional cases of osteonecrosis, of presumed mechanical origin and presenting a diffuse BME pattern [20,27,58]. Like in the hip, collapsed epiphyseal osteonecrosis may represent the common end point of two different initiating insults: vascular failure in AVN and mechanical injury in SONK [3,59].

MR images frequently show SIFs in SONK (Fig. 2). Open-ended low signal intensity lines suggestive for SIFs are frequently observed within the subchondral bone in SONK [52,54]. These fractures have been reported in as many as 78% of cases in a series of patients with SONK [52]. This common feature between SONK and SIFs suggests that these lesions could be closely related, could share a common origin with opposed outcomes, or could simply represent a spectrum of severity of the same entity [52]. The concurrent observation on MR imaging studies of typical SONK lesions in a medial femoral condyle, and of typical SIF in the adjacent medial tibial plateau, also lends credence to the postulate that SONK has a microtraumatic etiology [60].

MR imaging studies confirm the relationship between SONK and meniscal tears. The coexistence of meniscal lesions in the same knee compartment as SONK lesions has been noted for years, suggesting that SONK may be secondary to altered mechanical stress [17,31,32,42,48]. Several MR imaging studies confirmed this association between SONK and meniscal tears [4,52]. The cardinal role of radial and root tears of the posterior horn of the medial meniscus in the pathogenesis of SONK and SIFs has been underlined (Fig. 3) [54].

Histology confirms the presence of microfractures within the subchondral bone adjacent to SONK lesions [17,43]. Yamamoto and Bullough [17] retrospectively reviewed the pathologic specimens in a series of operated patients with the clinical and pathologic diagnosis of SONK. They observed the association of a subchondral fracture (fracture line with associated callus, reactive cartilage, and granulation tissue running grossly parallel to the articular surface) underlying a limited subchondral area of necrosis, interposed between the fracture line and the subchondral bone plate, which mirrors the observations on high-resolution MR images (Fig. 2) [17]. They concluded that the minimal extent of the area of necrosis and its strict location in the vicinity of the fracture line suggest that the fracture was the primary event [17]. These findings parallel their previous observations of SIFs underlying the subchondral bone collapse in cases of femoral head necrosis of presumed mechanical origin and lacking the typical MR imaging features of AVN [61–63].

Treatment

First-line treatment of SONK is conservative, consisting of protected weight bearing and analgesics, to prevent subchondral bone collapse and extension of epiphyseal osteonecrosis. Conservative treatment associated with serial radiographic follow-up may be pursued for several months or years in small lesions. For larger lesions with persisting symptoms, progressive collapse of the femoral condyle surface, and subsequent varus deformity, surgical treatment is indicated [9,41,43,44]. The different options are tibial valgisation osteotomy to relieve load on the medial compartment and unicompartimental or most often total knee arthroplasty [64,65].

Other conditions presenting the bone marrow edema pattern

Several self-resolving entities share the BME pattern as prominent MR imaging feature (low signal intensity on T1-weighted and high signal intensity on T2-weighted images). Two groups of lesions are usually considered, paralleling the categorization used for transient BME lesions of the femoral head [66–70]. In the first group, lesions are characterized by the observation of an isolated BME pattern on MR images, frequently associated with osteoporosis on radiographs, or particular clinical symptoms: transient osteoporosis (TO), transient BME, and reflex sympathetic dystrophy (RSDS) may be considered. In the second group, the hallmark is the demonstration of SIFs on imaging studies.

Besides very close and sometimes indistinguishable imaging appearances (BME on MR images and regional uptake on bone scintigrams), these conditions also present indistinct and probably overlapping

margins for the clinician. In the French literature, TO and transient BME tend to be considered as particular expressions of the multifaceted disease called *algodystrophy* or RSDS [71,72]. In the English literature, several authors separate TO (and migratory osteoporosis) from RSDS on the basis of the prompt healing without joint or bone sequelae in TO, in opposition to the longer course and possible late sequelae in RSDS of the foot and knee [58,73].

The possible overlap between SIFs and lesions presenting isolated BME has also been emphasized. The hypothesis of an occult SIF (missed on MR images) as a possible underlying mechanism of transient BME of the femoral condyles has been proposed [16,48,74]. This close relationship is particularly evident in renal transplant patients, in whom transient BME may be observed, either isolated or in association with clinical findings suggestive of RSDS, or with MR imaging findings suggestive of SIFs [74–78].

These different diagnoses are probably often made arbitrarily according to the relative weight of clinical information (typical symptoms for RSDS); radiographic findings (transient osteoporosis); and MR imaging information (pure BME, image of SIFs). Regardless of the definite appellation of these entities, the most important challenge for the radiologist is early identification and distinction from irreversible lesions (SONK and AVN), because most patients with these transient conditions recover completely with conservative treatment.

Transient osteoporosis, transient bone marrow edema syndrome, and reflex sympathetic dystrophy syndrome

TO is a spontaneously resolutive painful joint condition characterized by development of a radiographically evident osteopenia within several weeks after the onset of symptoms. Clinical status and imaging studies usually return to normal within 6 to 12 months. Joint space remains normal on radiographs. If other joints are subsequently involved, the disease is then termed *regional migratory osteoporosis*. The knee is the second most commonly affected joint after the hip. MR images show isolated BME and CT may be helpful to detect this regional bone rarefaction (Fig. 5). The possible observation of skin or soft tissue changes underlines the likely relationship between TO and RSDS (Fig. 5) [71, 79,80].

Transient BME is described in patients presenting similar clinical symptoms as those seen in TO, positive bone scans, and the BME pattern on MR images, but who do not develop evident osteopenia

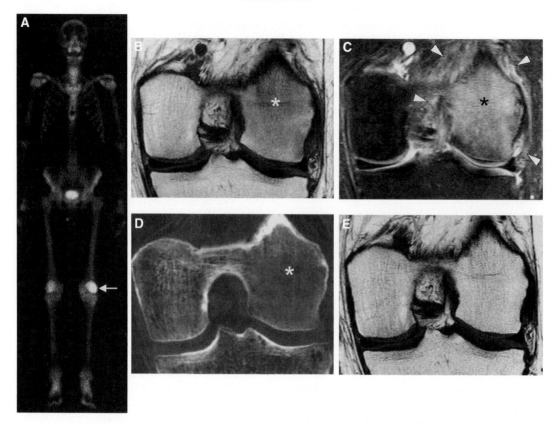

Fig. 5. Transient osteoporosis of the left knee in a 55-year-old man. (*A*) AP bone scintigram shows isotope uptake within the lateral femoral condyle (*arrow*). Coronal SE T1-weighted (*B*) and fat-saturated FSE T2-weighted (*C*) images show area of BME within the lateral femoral condyle (*). Note edema in adjacent soft tissues (*arrowheads in C*). (*D*) Reformatted coronal CT slice shows evident bone rarefaction and blurring of the trabecular network within the involved condyle (*). (*E*) A 6-month follow-up coronal SE T1-weighted image shows complete disappearance of BME.

on radiographs [81]. A migratory pattern of the BME has also been reported [82,83].

RSDS of the knee consists of exaggerated or inappropriate signs and symptoms following an injury or surgery, in general arthroscopy or arthrotomy, especially for total knee replacement [84–86]. The inappropriate symptoms include hypersensitivity, vasomotor disturbances, temperature changes, skin changes, impaired mobility, and sometimes progression to soft tissue contracture and atrophy. These symptoms typically respond to calcitonin or lower sympathetic blockade [84,87]. In some cases, the disease may be focal, limited to a knee compartment [84,88–90]. Radiographs may show patchy or diffuse osteopenia, but are nonspecific, because these findings may be observed as a result of disuse without superimposed RSDS. They may also remain normal, or show very subtle changes, especially in localized forms [72,91]. Bone scintigram is useful for the diagnosis in localized forms, but lacks specificity in

most cases, showing diffuse uptake [86,89]. MR images show soft tissue changes consisting of periarticular and subcutaneous edema and enhancement after contrast injection, skin thickening, muscle atrophy in late stages, and frequently joint effusion [73,92]. Observation of BME is less consistent [73,93]. Rankine et al [94] emphasize the role of the STIR sequence because changes may be discrete on T1- and T2-weighted images. In the experience of the authors and others, imaging findings parallel those seen in RSDS of the hip, consisting of BME seen within the involved femoral condyle on MR images, at least in early stages of the disease, sometimes showing a migratory pattern, and associated with regional bone rarefaction on concurrent radiographs [52,81,95].

Subchondral insufficiency fracture

SIFs are observed in epiphyses that are submitted to normal biomechanical stress but are weakened by nontumorous diseases, most frequently by osteoporosis; less frequent predisposing factors are osteomalacia, osteogenesis imperfecta, and hyperparathyroidism [17]. Most SIFs resolve after conservative therapy [62,96]. Reports have described patients who developed subchondral collapse and were treated surgically, however, which emphasizes the need for an early diagnosis of these lesions and elimination of weight bearing as conservative treatment [48,62].

Over the last 15 years, this entity has been widely described in the hip, thanks to the widespread use of high-resolution MR images obtained using surface coils [25,67,97–99]. This entity has also been clearly identified in knee epiphyses, presenting a typical MR imaging appearance [48,74]. The consistent observation is a low signal intensity line running almost parallel to the subchondral bone plate at a variable distance from the epiphyseal surface, representing the fracture plane, and surrounded by BME (Figs. 6 and 7) [48,52,57]. This appearance parallels that of occult posttraumatic fractures [100–102]. These lines are best seen on T2-weighted and fat-saturated intermediate-weighted MR images. The lesion shows predilection for the medial femoral condyle and for

its central weight-bearing area, but may involve other knee epiphyses [57,103]. Radiographs and CT are normal early in the clinical course, but may be helpful later for the confident identification of the lesion, which appears as a subchondral area of ill-delimited or more linear sclerosis (Figs. 6 and 7).

Differential diagnosis between spontaneous osteonecrosis of the knee and self-resolving lesions presenting the bone marrow edema pattern

When facing spontaneous painful epiphyseal lesions of the knee presenting the BME pattern on MR images, the critical point is to be able as early as possible to differentiate lesions that resolve under conservative therapy (SIFs, TO, transient BME, RSDS) from SONK, which may evolve to irreparable epiphyseal collapse and joint destruction.

The prognostic value of MR imaging in these spontaneous painful femoral condyle lesions has been studied by correlation of the initial MR images with the spontaneous long-term outcome of involved epiphyses, consisting of either epiphyseal collapse, pathognomonic for SONK, or complete resolution [52,104]. Most important prognostic information is

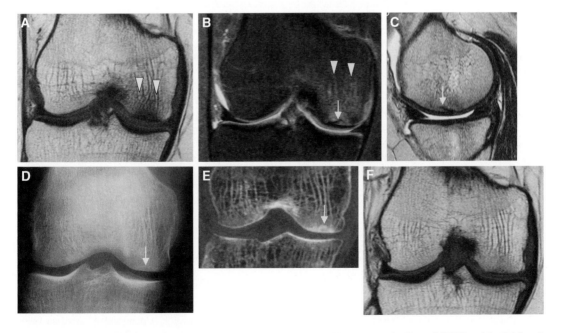

Fig. 6. Subchondral insufficiency fracture (SIF) in a 59-year-old woman with osteoporosis. Coronal SE T1-weighted (*A*) and fat-saturated FSE T2-weighted (*B*) images show BME within the medial femoral condyle (*arrowheads*). A low signal intensity line typical for a SIF is seen on the T2-weighted image (*arrow in B*). (*C*) Sagittal T2-weighted image shows the fracture line (*arrow*). Concurrent AP radiograph (*D*) and reformatted coronal CT slice (*E*) show the linear sclerosis of SIF (*arrows*). (*F*) A 6-month follow-up coronal T1-weighted image shows complete disappearance of subchondral marrow abnormalities.

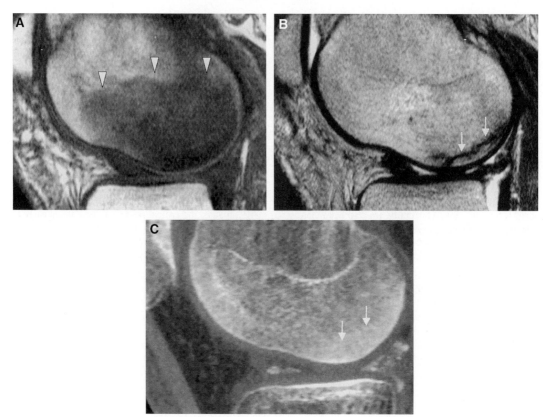

Fig. 7. SIF in an 83-year-old man with osteomalacia. Sagittal SE T1-weighted (*A*) and FSE T2-weighted (*B*) images show BME within the lateral femoral condyle (*arrowheads in A*). An extensive line of low signal intensity typical for a SIF is seen on the T2-weighted image (*arrows in B*). (*C*) Reformatted sagittal CT slice shows irregular sclerosis in this area (*arrows*) (note meniscocalcinosis).

provided by careful analysis of the subchondral bone marrow on T2-weighted (or enhanced T1-weighted) MR images. Most specific MR imaging features for the diagnosis of SONK are the presence and thickness of subchondral areas of low signal intensity (areas thicker than 4 mm, longer than 14 mm, or presenting a surface area superior to 3 cm^2 are pathognomonic for SONK) [52,104]. Lesions without this feature or with very limited subchondral low signal intensity areas tend to be reversible. If present on initial MR imaging studies, epiphyseal contours deformity and subchondral fluid-like signal intensity fracture clefts are also suggestive of the diagnosis of SONK. On the contrary, the presence of SIFs and the extent of BME have no value to differentiate SONK from transient lesions (Table 2) [52,104].

Table 2
Differential diagnosis between spontaneous osteonecrosis of the knee and self-resolving conditions presenting the bone marrow edema pattern

MR imaging features	SONK	Self-resolving BME
BME	+ + +	+ + +
Subchondral areas of low SI on T2 (necrosis)	+ + (Very specific if extensive)	+ (Rare and of limited thickness)
Lines of low SI on T1 and T2 at distance from surface (fractures)	+ +	+ + (SIF) ± (Rarely in TO, RSDS)
Contour deformity	+	± (Rare and subtle)
Subchondral fluid-like high SI on T2 (fracture cleft)	± (Rare but specific)	−

Abbreviations: −, absent; ±, rare; +, occasional; + +, frequent; + + +, always present.

If definite lesion categorization and prognosis of its spontaneous outcome remain uncertain on the basis of initial MR imaging findings, repeat MR imaging investigations a few months after the initial study may contribute to assess the outcome of the lesion under conservative treatment. Partial or complete regression indicates a likely favorable outcome; extension of subchondral low signal areas on T2-weighted images or appearance of other pejorative MR imaging fea-

tures (cortical deformity, subchondral fracture cleft) points toward an unfavorable evolution.

Particular setting: epiphyseal lesions observed after meniscectomy

Multiple MR imaging studies performed after meniscectomy reported the observation of a wide spectrum of lesions including self-resolving limited

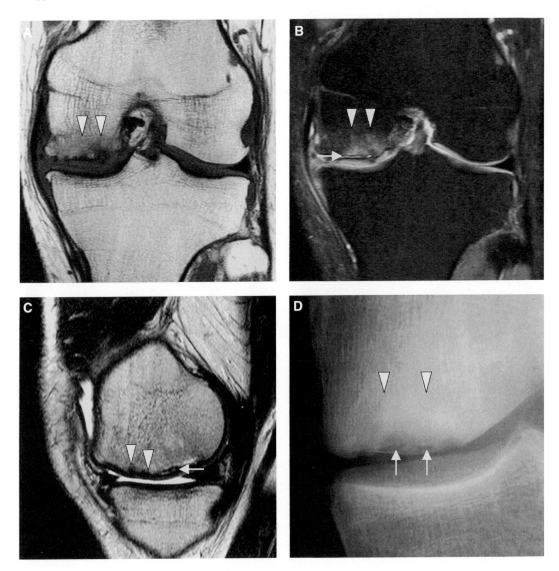

Fig. 8. SONK in a 44-year-old man 2 years after meniscectomy. Coronal SE T1-weighted (*A*) and fat-saturated FSE T2-weighted (*B*) images of the left knee show BME within the medial femoral condyle (*arrowheads*). A subchondral fracture cleft of fluid-like high signal intensity is seen (*arrow in B*). Medial segment of the medial meniscus is not visualized. (*C*) Sagittal T2-weighted FSE image shows subchondral areas of low signal intensity (*arrowheads*) and linear component of high signal intensity (fracture cleft) (*arrow*), suggestive for SONK. (*D*) Follow-up AP radiograph shows collapsed surface of the medial femoral condyle (*arrows*) surrounded by bone sclerosis (*arrowheads*), which confirms SONK.

BME, SIFs, and typical cases of SONK (Fig. 8) [49,105–109]. Most of these lesions are infraclinical: reactional marrow signal changes are seen on MR images in as many as one third of asymptomatic patients after meniscectomy [107]. Most of these lesions are completely resolutive under conservative treatment [107,110]. The likely hypothesis to explain these lesions is that altered knee biomechanics after meniscectomy may induce mechanical stress and cause microfractures within the subchondral bone, with in rare cases potential subsequent evolution to epiphyseal osteonecrosis.

Caveat osteoarthritis

All previous statements on the value of MR images for lesion identification, differential diagnosis, and prognosis apply to lesions that show normal articular cartilage. If cartilage lesions are present, the previously described semiology is biased, prognosis is uncertain, and does not depend on associated subchondral changes.

In degenerative diseases, cartilage lesions are indeed associated with a wide variety of bone marrow signal alterations, which may present the BME pattern and even simulate necrosis. The most common observation in OA is the BME pattern underlying the cartilage defects, which seems to be more frequent in symptomatic patients [111,112]. Histology has shown that this BME pattern reflects various histologic changes (fibrosis, cellular necrosis, abnormal bone trabeculae) and that edema is only a minor contributor to these changes [113]. Beside this BME pattern, subchondral hemispheric areas of low signal intensity on T1-weighted and intermediate or low signal–intensity on T2-weighted images may be seen in knees with OA, resembling the subchondral low signal intensity areas seen in SONK, and corresponding at histology to fibrosis, trabecular thickening and sometimes neocartilage, cysts, and even necrotic lacunae in some trabeculae [112]. These authors recommend examination of adjacent cartilage to look for significant diffuse or focal cartilage loss before making the diagnosis of SONK or of other conditions presenting the BME pattern (which should not be made if severe cartilage loss is seen) [112].

Summary

MR imaging is the technique of choice for the work-up of spontaneous painful lesions of the knee epiphyses. Two main categories of lesions have to be considered on T1-weighted images: AVN, presenting a geographic pattern, and lesions presenting the BME pattern. T1-weighted images are often sufficient to distinguish these two main categories. The latter category includes different entities referred to as SONK and self-resolving conditions, which in most cases may be differentiated by the study of the subchondral bone marrow area on high-resolution T2-weighted images. Patient demographics and analysis of radiographs or CT also contribute to this differential diagnosis.

Behind definite appellation and categorization of these lesions, the distinction between lesions that resolve spontaneously under conservative treatment from those that may evolve to epiphyseal collapse (AVN and SONK) is possible in most cases. This ability of MR imaging for early detection and prognosis of epiphyseal lesions of the knee affords the opportunity of instituting early individualized treatment.

References

[1] Glimcher MJ, Kenzora JE. The biology of osteonecrosis of the human femoral head and its clinical implications: II. The pathological changes in the femoral head as an organ and in the hip joint. Clin Orthop 1979;139:283–312.

[2] Robbins SL, Cotran RS, Kumar V. Cellular injury and adaptation. In: Robbins SL, Cotran RS, Kumar V, editors. Pathologic basis of disease. Philadelphia: WB Saunders; 1984. p. 1–39.

[3] Vande Berg BC, Malghem J, Lecouvet FE, et al. Magnetic resonance imaging and differential diagnosis of epiphyseal osteonecrosis. Semin Musculoskelet Radiol 2001;5:57–67.

[4] Narvaez J, Narvaez JA, Rodriguez-Moreno J, et al. Osteonecrosis of the knee: differences among idiopathic and secondary types. Rheumatology (Oxford) 2000;39:982–9.

[5] Vande Berg BC, Malghem J, Lecouvet FE, et al. Fat conversion of femoral marrow in glucocorticoid-treated patients: a cross-sectional and longitudinal study with magnetic resonance imaging. Arthritis Rheum 1999;42:1405–11.

[6] Isono SS, Woolson ST, Schurman DJ. Total joint arthroplasty for steroid-induced osteonecrosis in cardiac transplant patients. Clin Orthop 1987;217: 201–8.

[7] Havel PE, Ebraheim NA, Jackson WT. Steroid-induced bilateral avascular necrosis of the lateral femoral condyles: a case report. Clin Orthop 1989; 243:166–8.

[8] Kelman GJ, Williams GW, Colwell Jr CW, et al. Steroid-related osteonecrosis of the knee: two case

reports and a literature review. Clin Orthop 1990; 257:171–6.

[9] Motohashi M, Morii T, Koshino T. Clinical course and roentgenographic changes of osteonecrosis in the femoral condyle under conservative treatment. Clin Orthop 1991;266:156–61.

[10] Stovall Jr D, Young TR. Avascular necrosis of the medial femoral condyle in HIV-infected patients. Am J Orthop 1995;24:71–3.

[11] Bahebeck J, Atangana R, Techa A, et al. Relative rates and features of musculoskeletal complications in adult sicklers. Acta Orthop Belg 2004;70:107–11.

[12] Kenzora JE, Glimcher MJ. Accumulative cell stress: the multifactorial etiology of idiopathic osteonecrosis. Orthop Clin North Am 1975;16:669–79.

[13] Mont MA, Hungerford DS. Non-traumatic avascular necrosis of the femoral head. J Bone Joint Surg Am 1995;77:459–74.

[14] Chancelier MD, Helenon O, Page B, et al. Aseptic osteonecrosis of the knee induced by corticoids: MRI aspects. J Radiol 1992;73:191–201.

[15] Mankin HJ. Nontraumatic necrosis of bone (osteonecrosis). N Engl J Med 1992;326:1473–9.

[16] Malghem J, Le Hir P, Lecouvet F, et al. Ostéonécrose du genou. In: Bard H, Drapé JL, Goutallier D, et al, editors. GETROA Opus XXIV. Le genou traumatique et dégénératif. Montpellier: Sauramps Medical; 1997. p. 117–36.

[17] Yamamoto T, Bullough PG. Spontaneous osteonecrosis of the knee: the result of subchondral insufficiency fracture. J Bone Joint Surg Am 2000;82: 858–66.

[18] Matsuo K, Hirohata T, Sugioka Y, et al. Influence of alcohol intake, cigarette smoking, and occupational status on idiopathic osteonecrosis of the femoral head. Clin Orthop 1988;234:115–23.

[19] Guerra JJ, Steinberg ME. Distinguishing transient osteoporosis from avascular necrosis of the hip. J Bone Joint Surg Am 1995;77:616–24.

[20] Mitchell DG, Rao VM, Dalinka MK, et al. Femoral head avascular necrosis: correlation of MR imaging, radiographic staging, radionuclide imaging, and clinical findings. Radiology 1987;162:709–15.

[21] Mitchell DG, Joseph PM, Fallon M, et al. Chemical-shift MR imaging of the femoral head: an in vitro study of normal hips and hips with avascular necrosis. AJR Am J Roentgenol 1987;148:1159–64.

[22] Vande Berg BC, Malghem J, Labaisse MA, et al. MR imaging of avascular necrosis and transient marrow edema of the femoral head. Radiographics 1993;13: 501–20.

[23] Sugimoto H, Tanaka O, Ohsawa T. MR imaging of femoral head avascular necrosis with STIR sequence. Nippon Igaku Hoshasen Gakkai Zasshi 1989;49: 1067–9.

[24] Duda SH, Laniado M, Schick F, et al. The double-line sign of osteonecrosis: evaluation on chemical shift MR images. Eur J Radiol 1993;16:233–8.

[25] Watson RM, Roach NA, Dalinka MK. Avascular necrosis and bone marrow edema syndrome. Radiol Clin North Am 2004;42:207–19.

[26] Lang P, Jergesen HE, Moseley ME, et al. Avascular necrosis of the femoral head: high-field-strength MR imaging with histologic correlation. Radiology 1988; 169:517–24.

[27] Vande Berg BC, Malghem J, Labaisse MA, et al. Avascular necrosis of the hip: comparison of contrast-enhanced and nonenhanced MR imaging with histologic correlation: work in progress. Radiology 1992; 182:445–50.

[28] Mitchell DG, Steinberg ME, Dalinka MK, et al. Magnetic resonance imaging of the ischemic hip: alterations within the osteonecrotic, viable, and reactive zones. Clin Orthop 1989;244:60–77.

[29] Sakai T, Sugano N, Ohzono K, et al. MRI evaluation of steroid- or alcohol-related osteonecrosis of femoral condyle. Acta Orthop Scand 1998;69:598–602.

[30] Mont MA, Tomek IM, Hungerford DS. Core decompression for avascular necrosis of the distal femur: long term follow-up. Clin Orthop 1997;334:124–30.

[31] Ahlback S, Bauer GC, Bohne WH. Spontaneous osteonecrosis of the knee. Arthritis Rheum 1968;11: 705–33.

[32] Rozing PM, Insall J, Bohne WH. Spontaneous osteonecrosis of the knee. J Bone Joint Surg Am 1980;62:2–7.

[33] Aglietti P, Insall JN, Buzzi R, et al. Idiopathic osteonecrosis of the knee: aetiology, prognosis and treatment. J Bone Joint Surg Br 1983;65:588–97.

[34] al Rowaih A, Lindstrand A, Bjorkengren A, et al. Osteonecrosis of the knee: diagnosis and outcome in 40 patients. Acta Orthop Scand 1991;62:19–23.

[35] Houpt JB, Alpert B, Lotem M, et al. Spontaneous osteonecrosis of the medial tibial plateau. J Rheumatol 1982;9:81–90.

[36] Ecker ML, Lotke PA. Osteonecrosis of the medial part of the tibial plateau. J Bone Joint Surg Am 1995; 77:596–601.

[37] Ahuja SC, Bullough PG. Osteonecrosis of the knee: a clinicopathological study in twenty-eight patients. J Bone Joint Surg Am 1978;60:191–7.

[38] Marmor L. Osteonecrosis of the knee: medial and lateral involvement. Clin Orthop 1984;185:195–6.

[39] Lotke PA, Ecker ML. Osteonecrosis-like syndrome of the medial tibial plateau. Clin Orthop 1983;176: 148–53.

[40] Lotke PA, Nelson CL, Lonner JH. Spontaneous osteonecrosis of the knee: tibial plateaus. Orthop Clin North Am 2004;35:365–70.

[41] Lotke PA, Ecker ML, Alavi A. Painful knees in older patients: radionuclide diagnosis of possible osteonecrosis with spontaneous resolution. J Bone Joint Surg Am 1977;59:617–21.

[42] Norman A, Baker ND. Spontaneous osteonecrosis of the knee and medial meniscal tears. Radiology 1978; 129:653–6.

[43] Hernigou P. Idiopathic osteonecrosis of the internal femoral condyle: prognostic elements and role

of different treatments. Rev Rhum Ed Fr 1993;60: 203–11.

[44] Bullough PG, DiCarlo EF. Subchondral avascular necrosis: a common cause of arthritis. Ann Rheum Dis 1990;49:412–20.

[45] Pollack MS, Dalinka MK, Kressel HY, et al. Magnetic resonance imaging in the evaluation of suspected osteonecrosis of the knee. Skeletal Radiol 1987;16:121–7.

[46] Lotke PA, Ecker ML. Osteonecrosis of the knee. J Bone Joint Surg Am 1988;70:470–3.

[47] Muheim G, Bohne WH. Prognosis in spontaneous osteonecrosis of the knee: investigation by radionuclide scintimetry and radiography. J Bone Joint Surg Br 1970;52:605–12.

[48] Lafforgue P, Daumen-Legre V, Clairet D, et al. Insufficiency fractures of the medial femoral condyle. Rev Rhum Engl Ed 1996;63:262–9.

[49] Kursunioglu-Brahme S, Fox JM, Ferkel RD, et al. Osteonecrosis of the knee after arthroscopic surgery: diagnosis with MR imaging. Radiology 1991;178: 851–3.

[50] Lang P, Grampp S, Vahlensieck M, et al. Spontaneous osteonecrosis of the knee joint: MRT compared to CT, scintigraphy and histology. Rofo Fortschr Geb Rontgenstr Neuen Bildgeb Verfahr 1995;162:469–77.

[51] Bjorkengren AG, Al Rowaih A, Lindstrand A, et al. Spontaneous osteonecrosis of the knee: value of MR imaging in determining prognosis. AJR Am J Roentgenol 1990;154:331–6.

[52] Lecouvet FE, Vande Berg BC, Maldague BE, et al. Early irreversible osteonecrosis versus transient lesions of the femoral condyles: prognostic value of subchondral bone and marrow changes on MR imaging. AJR Am J Roentgenol 1998;170:71–7.

[53] Bootsveld K, Siewert B, Reiser M, et al. Spontaneous necrosis of the femoral condyle: new findings in T2- weighted spin-echo sequences and gradient-echo studies. Rofo Fortschr Geb Rontgenstr Neuen Bildgeb Verfahr 1992;156:360–4.

[54] Yao L, Stanczak J, Boutin RD. Presumptive subarticular stress reactions of the knee: MRI detection and association with meniscal tear patterns. Skeletal Radiol 2004;33:260–4.

[55] Hall FM. Osteonecrosis of the knee and medial meniscal tears. Radiology 1979;133(3 Pt 1):828–9.

[56] Zanetti M, Romero J, Dambacher MA, et al. Osteonecrosis diagnosed on MR images of the knee: relationship to reduced bone mineral density determined by high resolution peripheral quantitative CT. Acta Radiol 2003;44:525–31.

[57] Ramnath RR, Kattapuram SV. MR appearance of SONK-like subchondral abnormalities in the adult knee: SONK redefined. Skeletal Radiol 2004;33: 575–81.

[58] Froberg PK, Braunstein EM, Buckwalter KA. Osteonecrosis, transient osteoporosis, and transient bone marrow edema: current concepts. Radiol Clin North Am 1996;34:273–91.

[59] Mitchell DG. Using MR imaging to probe the pathophysiology of osteonecrosis. Radiology 1989;171: 25–6.

[60] Sokoloff RM, Farooki S, Resnick D. Spontaneous osteonecrosis of the knee associated with ipsilateral tibial plateau stress fracture: report of two patients and review of the literature. Skeletal Radiol 2001; 30:53–6.

[61] Yamamoto T, Bullough PG. Subchondral insufficiency fracture of the femoral head: a differential diagnosis in acute onset of coxarthrosis in the elderly. Arthritis Rheum 1999;42:2719–23.

[62] Yamamoto T, Bullough PG. Subchondral insufficiency fracture of the femoral head and medial femoral condyle. Skeletal Radiol 2000;29:40–4.

[63] Yamamoto T, Schneider R, Bullough PG. Insufficiency subchondral fracture of the femoral head. Am J Surg Pathol 2000;24:464–8.

[64] al Rowaih A, Bjorkengren A, Egund N, et al. Size of osteonecrosis of the knee. Clin Orthop 1993;287: 68–75.

[65] Soucacos PN, Xenakis TH, Beris AE, et al. Idiopathic osteonecrosis of the medial femoral condyle: classification and treatment. Clin Orthop 1997;341: 82–9.

[66] Hayes CW, Conway WF, Daniel WW. MR imaging of bone marrow edema pattern: transient osteoporosis, transient bone marrow edema syndrome, or osteonecrosis. Radiographics 1993;13:1001–11.

[67] Vande Berg BC, Malghem J, Maldague B. MR imaging of equivocal femoral head lesions: diagnosis or prognosis. Radiology 1997;203:290–1.

[68] Bloem JL. Transient osteoporosis of the hip: MR imaging. Radiology 1988;167:753–5.

[69] Wilson AJ, Murphy WA, Hardy DC, et al. Transient osteoporosis: transient bone marrow edema? Radiology 1988;167:757–60.

[70] Trepman E, King TV. Transient osteoporosis of the hip misdiagnosed as osteonecrosis on magnetic resonance imaging. Orthop Rev 1992;21:1089–98.

[71] Lequesne M. Transient osteoporosis of the hip: a nontraumatic variety of Südeck's atrophy. Ann Rheum Dis 1968;27:463–71.

[72] Doury P. Bone-marrow oedema, transient osteoporosis, and algodystrophy [letter; comment]. J Bone Joint Surg Br 1994;76:993–4.

[73] Schweitzer ME, Mandel S, Schwartzman RJ, et al. Reflex sympathetic dystrophy revisited: MR imaging findings before and after infusion of contrast material. Radiology 1995;195:211–4.

[74] Vande Berg BC, Malghem J, Goffin EJ, et al. Transient epiphyseal lesions in renal transplant recipients: presumed insufficiency stress fractures. Radiology 1994;191:403–7.

[75] Coates PT, Tie M, Russ GR, et al. Transient bone marrow edema in renal transplantation: a distinct post-transplantation syndrome with a characteristic MRI appearance. Am J Transplant 2002;2: 467–70.

[76] Naredo SE, Balsa CA, Sanz GA, et al. Leg bone pain syndrome due to cyclosporine in a renal transplant patient. Clin Exp Rheumatol 1994;12:653–6.

[77] Munoz-Gomez J, Collado A, Gratacos J, et al. Reflex sympathetic dystrophy syndrome of the lower limbs in renal transplant patients treated with cyclosporin A. Arthritis Rheum 1991;34:625–30.

[78] Bouteiller G, Lloveras JJ, Condouret J, et al. Painful polyarticular syndrome probably induced by cyclosporin in three patients with a kidney transplant and one with a heart transplant. Rev Rhum Mal Osteoartic 1989;56:753–5.

[79] Swezey RL. Transient osteoporosis of the hip, foot and knee. Arthritis Rheum 1970;13:858–68.

[80] Lakhanpal S, Ginsburg WW, Luthra HS, et al. Transient regional osteoporosis: a study of 56 cases and review of the literature. Ann Intern Med 1987; 106:444–50.

[81] Fertakos RJ, Swayne LC, Colston WC. Three-phase bone imaging in bone marrow edema of the knee. Clin Nucl Med 1995;20:587–90.

[82] Moosikasuwan JB, Miller TT, Math K, et al. Shifting bone marrow edema of the knee. Skeletal Radiol 2004;33:380–5.

[83] Papadopoulos EC, Papagelopoulos PJ, Kaseta M, et al. Bone marrow edema syndrome of the knee: a case report and review of the literature. Knee 2003; 10:295–302.

[84] Fournier RS, Holder LE. Reflex sympathetic dystrophy: diagnostic controversies. Semin Nucl Med 1998;28:116–23.

[85] O'Brien SJ, Ngeow J, Gibney MA, et al. Reflex sympathetic dystrophy of the knee: causes, diagnosis, and treatment. Am J Sports Med 1995;23:655–9.

[86] Cuartero-Plaza A, Martinez-Miralles E, Benito-Ruiz P, et al. Abnormal bone scintigraphy and silent radiography in localized reflex sympathetic dystrophy syndrome. Eur J Nucl Med 1992;19:330–3.

[87] Ogilvie-Harris DJ, Roscoe M. Reflex sympathetic dystrophy of the knee. J Bone Joint Surg Br 1987; 69:804–6.

[88] Lequesne M, Kerboull M, Bensasson M, et al. Partial decalcifying algodystrophy. Rev Rhum Mal Osteoartic 1979;46:111–21.

[89] Doury P. Partial, limited and subradiologic atypical forms of algodystrophy. Rev Rhum Mal Osteoartic 1982;49:781–6.

[90] Doury P, Pattin S, Eulry F, et al. Algodystrophy of the knee: apropos of a series of 125 cases. Rev Rhum Mal Osteoartic 1987;54:655–9.

[91] Malhotra R, Dhingra SS, Padhy AK, et al. Reflex sympathetic dystrophy of the patello-femoral joint: diagnosis and relevance. Clin Nucl Med 1995;20: 1058–60.

[92] Graif M, Schweitzer ME, Marks B, et al. Synovial effusion in reflex sympathetic dystrophy: an additional sign for diagnosis and staging. Skeletal Radiol 1998;27:262–5.

[93] Koch E, Hofer HO, Sialer G, et al. Failure of MR imaging to detect reflex sympathetic dystrophy of the extremities. AJR Am J Roentgenol 1991;156:113–5.

[94] Rankine JJ, Smith FW, Scotland TR. Case report: short tau inversion recovery (STIR) sequence MRI appearances of reflex sympathetic dystrophy. Clin Radiol 1995;50:188–90.

[95] Hauzeur JP, Hanquinet S, Gevenois PA, et al. Study of magnetic resonance imaging in transient osteoporosis of the hip. J Rheumatol 1991;18:1211–7.

[96] Bangil M, Soubrier M, Dubost JJ, et al. Subchondral insufficiency fracture of the femoral head. Rev Rhum Engl Ed 1996;63:859–61.

[97] Rafii M, Mitnick H, Klug J, et al. Insufficiency fracture of the femoral head: MR imaging in three patients. AJR Am J Roentgenol 1997;168:159–63.

[98] Grignon B, Pere P, Regent D, et al. Les fractures par insuffisance osseuse de l'épiphyse fémorale supérieure. J Radiol 1990;71:525–30.

[99] Kopecky KK, Braunstein EM, Brandt KD, et al. Apparent avascular necrosis of the hip: appearance and spontaneous resolution of MR findings in renal allograft recipients. Radiology 1991;179:523–7.

[100] Vellet AD, Marks PH, Fowler PJ, et al. Occult posttraumatic osteochondral lesions of the knee: prevalence, classification, and short-term sequelae evaluated with MR imaging. Radiology 1991;178: 271–6.

[101] Mink JH, Deutsch AL. Occult cartilage and bone injuries of the knee: detection, classification, and assessment with MR imaging. Radiology 1989;170(3 Pt 1): 823–9.

[102] Yao L, Lee JK. Occult intraosseous fracture: detection with MR imaging. Radiology 1988;167:749–51.

[103] Le Gars L, Savy JM, Orcel P, et al. Osteonecrosis-like syndrome of the medial tibial plateau can be due to a stress fracture: MR findings in 13 patients. Rev Rhum Engl Ed 1999;66:323–30.

[104] Le Hir P, Larédo JD, Zeitoun F, et al. "Ostéonécrose" du condyle fémoral: le syndrome de contusion spontanée et l'ostéonécrose ischémique: savoir faire en radiologie ostéo-articulaire. Montpellier: Sauramps Medical; 2000. p. 135–49.

[105] Muscolo DL, Costa-Paz M, Makino A, et al. Osteonecrosis of the knee following arthroscopic meniscectomy in patients over 50-years old. Arthroscopy 1996;12:273–9.

[106] Prues-Latour V, Bonvin JC, Fritschy D. Nine cases of osteonecrosis in elderly patients following arthroscopic meniscectomy. Knee Surg Sports Traumatol Arthrosc 1998;6:142–7.

[107] Kobayashi Y, Kimura M, Higuchi H, et al. Juxtaarticular bone marrow signal changes on magnetic resonance imaging following arthroscopic meniscectomy. Arthroscopy 2002;18:238–45.

[108] Johnson TC, Evans JA, Gilley JA, et al. Osteonecrosis of the knee after arthroscopic surgery for meniscal tears and chondral lesions. Arthroscopy 2000;16:254–61.

[109] Nakamura N, Horibe S, Nakamura S, et al. Sub-

chondral microfracture of the knee without osteonecrosis after arthroscopic medial meniscectomy.
Arthroscopy 2002;18:538–41.

[110] Santori N, Condello V, Adriani E, et al. Osteonecrosis
after arthroscopic medial meniscectomy. Arthroscopy
1995;11:220–4.

[111] Felson DT, Chaisson CE, Hill CL, et al. The
association of bone marrow lesions with pain in knee
osteoarthritis. Ann Intern Med 2001;134:541–9.

[112] Bergman AG, Willen HK, Lindstrand AL, et al.
Osteoarthritis of the knee: correlation of subchondral MR signal abnormalities with histopathologic
and radiographic features. Skeletal Radiol 1994;23:
445–8.

[113] Zanetti M, Bruder E, Romero J, et al. Bone marrow
edema pattern in osteoarthritic knees: correlation
between MR imaging and histologic findings. Radiology 2000;215:835–40.

ELSEVIER
SAUNDERS

Radiol Clin N Am 43 (2005) 673 – 681

RADIOLOGIC
CLINICS
of North America

High- Versus Low-Field MR Imaging

Thierry Tavernier, MD[a], Anne Cotten, MD[b],*

[a]Imagerie Médicale, Clinique de la Sauvegarde, Lyon, France
[b]Service de Radiologie Ostéo-Articulaire, Hôpital Roger Salengro, Boulevard du Pr. J. Leclercq, Lille 59037, France

The role of MR imaging as a noninvasive technique in the detection and evaluation of musculoskeletal diseases is unquestionable, because MR imaging has the advantages of the combined evaluation of bones, ligaments, and soft tissue. Most of the studies reported in the literature are based on high-field MR imaging. Low-field MR imaging magnets are defined by a field-strength inferior to 0.5 T. They consist of open-configured whole-body MR imaging systems or dedicated extremities scanners. Initial studies performed with low-field-strength have reported unsatisfactory results in the assessment of the musculoskeletal system [1,2]. Recent improvements, however, have generated a renewed interest in low-field-strength MR imaging [3,4]. Indeed, the currently increasing use of dedicated low-field MR imaging systems in orthopedic applications results from a reported equivalent reliability compared with conventional high-field scanners. Only a few clinical studies have compared the diagnostic performance of low- and high-field-strength magnets in the diagnosis of joint disorders, however, mainly focusing on ligamentous and meniscal trauma of the knee [4–6]. This article presents the principal applications and results published in the literature.

Advantages of low-field systems

In the age of cost containment and urgent reduction in health care expenditures, low-field systems may represent a cost-effective alternative for MR imaging of musculoskeletal diseases, because the costs of buying, installing, and maintaining a low-field system are much less expensive than those of a high-field system. This is mainly because of the lower purchase price, cheaper maintenance contract, less floor space required, lower electricity consumption, lower number of technicians required, no need for magnetic shielding, and no need for cryogens [4]. These features explain why many studies concerning MR imaging at low field have been reported from countries with government-controlled health economies [4]. Other advantages of low-field systems are a higher comfort for claustrophobic patients and for children; the possibility of a central positioning of the area to explore because of the lateral displacement of the table in open systems; the possibility of MR imaging–guided interventions and kinematic studies; a higher tissue contrast, especially on T1-weighted sequences (Fig. 1); less metallic artifacts in patients who have undergone an operation (Fig. 2) [7]; and the absence of phase artifacts from pulsatile blood flow, especially at the knee [4]. Finally, stray fields of several dedicated low-field scanners may be so small that cardiac pacemakers and implantable cardioverter defibrillators may be not affected during imaging of the extremity.

Limits of low-field systems

The main limitations of low-field MR imaging result from a lower signal-to-noise ratio, which has to be compensated for by increasing the section thickness; reducing the in-plane resolution; increasing the number of acquisitions (and consecutively the acquisition time); and decreasing the bandwidth [8].

* Corresponding author.
E-mail address: acotten@chru-lille.fr (A. Cotten).

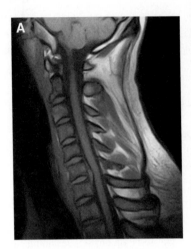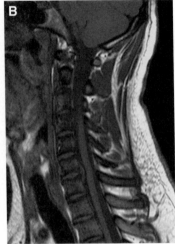

Fig. 1. Sagittal spin echo T1-weighted images of the cervical spine obtained at 0.3 T (*A*) and 1.5 T (*B*).

Furthermore, low-field scanners lack the capability of using sophisticated techniques, which require a strong and homogeneous magnetic field or a powerful gradient system. Recent improvements, however, including better magnet homogeneity, improved receiving coil technology [9], and dedicated pulse sequences have generated a renewed interest in low-field-strength MR imaging [3,4]. The usual sequences used for the assessment of the musculoskeletal system can be applied to the low-field (fast or turbo spin echo (TSE) sequences, gradient echo sequences, and STIR sequences).

Limitations in signal-to-noise ratio and in homogeneous static magnetic field also result in the inability to achieve fat-saturated imaging in several systems (≤ 0.2 T) and no reliable distinction between fat and contrast media [10]. Recently, three-point Dixon chemical shift imaging has been applied to low-field magnets allowing a reliable and homogeneous fat suppression (Figs. 2C, 3, and 4). In a single acquisition, water and fat are separated and static field inhomogeneities are corrected [11,12]. This imaging has been tested successfully in the musculoskeletal field, especially for acute bone fractures [13], cartilage evaluation in the knee joint, and for spine evaluation [8,10,11].

Knee

Meniscal and cruciate ligament tears

A limited number of clinical studies have directly compared the accuracies of different high- and low-field MR imaging systems for meniscal and anterior cruciate ligament tears [3–6]. The most recent studies showed comparable results for high- and low-fields, despite the difference in the image quality. Other investigators, who compared the diagnostic accuracies of low-field or ultra-low-field MR imaging (0.04–0.2 T) and diagnostic arthroscopy reported detection rates similar to those from large studies on high-field MR imaging diagnosis of meniscal and ligamentous knee joint injuries (Figs. 4 and 5) [2,14,15].

If diagnostically accurate images of the knee can be obtained with low-field MR imaging, the radiologist has to keep in mind that the level of confidence in decision making has been reported significantly superior with high-field-strength imaging, probably reflecting their higher conspicuity of lesions from high-field-strength units. This may signify a higher number of investigations with equivocal findings with low-field units, which might lead to second examination on high-field units [16].

Cartilage

Several clinical studies on low-field MR imaging of articular cartilage have suggested substantial limitations of this technique. Sensitivities of 25% to 75% for the detection of grade 2 cartilage lesions and 60% to 73% for grade 3 lesions using a 0.2-T dedicated MR imaging system have been reported [14,17,18]. Lower diagnostic accuracy of three-dimensional pulse sequences at 0.18 T compared with those obtained at high field has also been described. Compared with arthroscopy, however,

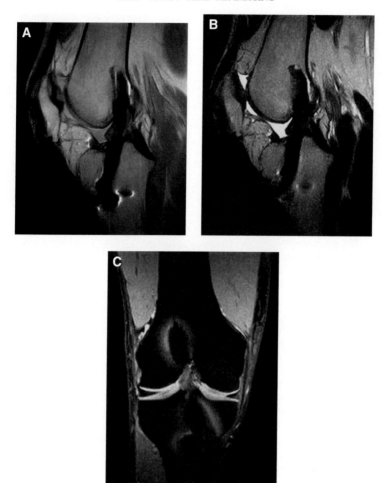

Fig. 2. Sagittal fast spin echo proton density-weighted (*A*) and T2-weighted (*B*) images and coronal Dixon gradient echo T2-weighted image with magnetization transfer contrast (*C*) of a ligamentoplasty of the knee (images obtained at 0.3 T).

Bredella et al [11] reported an overall sensitivity of 80% and a specificity of 73% using a combination of axial and sagittal planes of three-point Dixon sequence. The sensitivity and specificity was lower in detecting early stages of chondromalacia (grade 1) and a tendency of early cartilage abnormalities to be overestimated with the three-point Dixon sequence was noted in this study (Fig. 6). MR arthrography may also be used at low-field strength and has been reported as helpful in the diagnosis of chondromalacia patella [18].

Shoulder

There has been only a limited number of studies available dealing with low-field MR imaging and MR arthrography of the glenohumeral joint. They reveal that, despite minor image quality in comparison with high-field imaging, low-field systems may allow sufficient evaluation of intra-articular and extra-articular structures in the detection of major abnormalities, such as glenohumeral instability or rotator cuff disease [19]. Open-configured MR imaging units offer the possibility for performing MR imaging–guided arthrography of the shoulder. Petersilge et al [20] first described this technique using an open C-arm scanner with a vertically oriented magnetic field so that MR arthrography may be performed in one setting.

Rotator cuff

Shellock et al [21] in their study on unenhanced low-field imaging of 47 patients achieved a sensi-

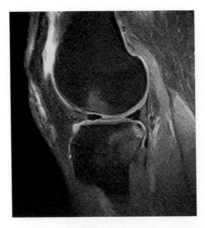

Fig. 3. Sagittal Dixon TSE proton density-weighted image of bone bruises associated with an anterior cruciate ligament tear (image obtained at a 0.3 T).

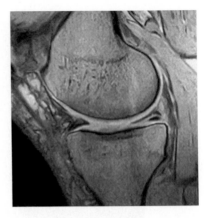

Fig. 5. Sagittal gradient echo T2-weighted image of a meniscal tear (image obtained at 0.3 T).

tivity and specificity of 89% and 100%, respectively, for the detection of rotator cuff tears. Merl et al [22] found no significant differences between unenhanced high-field and low-field imaging in diagnosing shoulder impingement and tears of the rotator cuff. In a study on 82 patients with surgical correlation, low-field MR arthrography achieved a sensitivity of 97% and a specificity of 100% in diagnosing complete tears of the rotator cuff [19]. Excellent or substantial interobserver agreement has also been reported with this technique [23]. The depiction of small tears of the rotator cuff may be influenced,

however, by the ability of the machine to perform Dixon or fat-saturated sequences (Figs. 7 and 8).

Instability

There are few reports dealing with the accuracy of low-field imaging with respect to glenohumeral instability. No significant discrepancies between unenhanced high-field and low-field imaging have been reported [22,24]. Studies on unenhanced low-field systems found a sensibility and specificity for labrum pathology of 77% and 91% and 67% and 95%, respectively [21,22,24], and for capsular lesions

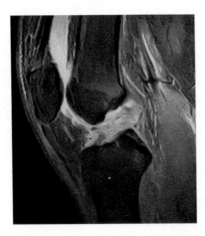

Fig. 4. Sagittal Dixon TSE proton density-weighted image of a torn anterior cruciate ligament (image obtained at a 0.3 T).

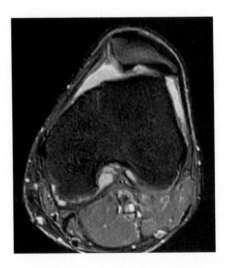

Fig. 6. Axial Dixon TSE proton density-weighted image showing a cartilage lesion of the lateral patellar facet and a trochlear dysplasia (image obtained at 0.3 T).

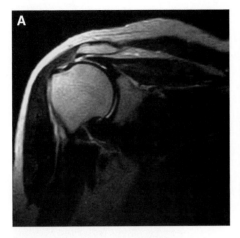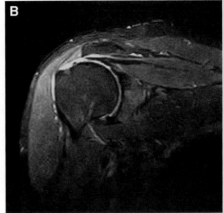

Fig. 7. Coronal TSE proton density-weighted (*A*) and Dixon TSE proton density-weighted (*B*) images showing a complete tear of the supraspinatus tendon (images obtained at 0.3 T).

of 63% and 80%, respectively [24]. In one study with direct comparison of low-field with high-field MR arthrography of the shoulder, sensibility and specificity were 100% and 93%, respectively, for both systems.

The performance characteristics of unenhanced low-field MR imaging have been reported inferior to those of high-field MR imaging for the diagnosis of superior labral anteroposterior lesions (accuracy of 67% versus 80%) [25]. Differences in imaging technique and in particular differences in spatial resolution may explain the lower sensitivity of low-field MR imaging in this study [25]. Three of the four

missed superior labral anteroposterior tears on low-field MR imaging, however, were arthroscopic type I superior labral anteroposterior tears. This type of superior labral anteroposterior tear represents a marked fraying of the superior labrum without labral detachment and some controversy exists as to whether this lesion is truly an injury or a degenerative lesion [25]. To the best of the authors' knowledge, there are no data of low-field MR arthrography with respect to the detection of superior labral anteroposterior lesions.

Trauma

The medical benefit of a low-field MR imaging system integrated in the traumatologic suite of a radiologic department has been reported as potentially useful because of the potential impact on the early initiation of the correct therapy, which may avoid additional diagnostic procedures and may reduce costs. Low-field MR imaging of the wrist when scaphoid fracture is suspected in patients with negative or equivocal radiographs may demonstrate scaphoid fractures and additional or simulating lesions, bone bruise of other carpal bones, and other carpal fractures [26,27]. These MR imaging features can alter the management of the patients with overall costs that have been reported to change little from conventional practice [27]. Herber et al [28] have also recommended low-field MR imaging of the ankle in children with persistent or unclear pain of the ankle joint and inconspicuous conventional radiograph because they found ligamental ruptures and

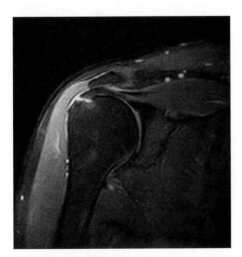

Fig. 8. Coronal STIR image showing a complete tear of the supraspinatus tendon (images obtained at 0.3 T).

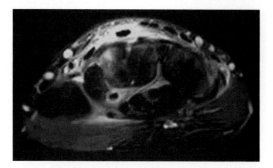

Fig. 9. Axial Dixon spin echo T1-weighted image performed after gadolinium administration of a rheumatoid wrist (image obtained at 0.3 T).

occult fractures changing the therapy in 64% of their patients.

Other usual MR imaging indications

In the assessment of rheumatologic diseases, the value of low-field MR imaging in the follow-up of patients with active rheumatoid arthritis has been reported (Fig. 9) [29–31], and in the early detection of sacroiliitis accompanying ankylosing spondylitis, because this technique can make more diagnoses of sacroiliitis than radiography [32]. In the study of Yu et al [32], low-field MR imaging was able to reveal early cartilage changes and bone marrow edema, which could not be found by either CT or radiography. Other entheses may also be assessed (Fig. 10).

Low-field systems have also been found helpful in the assessment of many other musculoskeletal

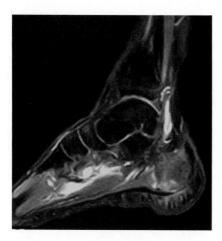

Fig. 10. Sagittal Dixon TSE proton density-weighted image showing enthesitis of the plantar aponeurosis in a patient with ankylosing spondylitis (image obtained at 0.3 T).

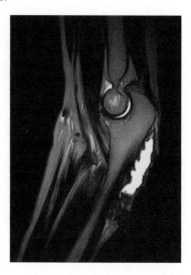

Fig. 11. Sagittal TSE T2-weighted image showing a subcutaneous bursitis of the elbow (image obtained at 0.3 T).

disorders, including osteonecrosis of the femoral head, osteomyelitis [33,34], chronic exertional compartment syndrome [35], lateral epicondylitis [36], full spectrum of disorders affecting the foot and ankle [37], and gamekeeper thumb (Fig. 11) [38].

Interventional procedures using low-field systems

Advantages and limits

One advantage of MR imaging versus fluoroscopy or CT guidance is the ability to have permanent access to the patient during imaging (hands-on technique) without exposure of the radiologist to radiation. Another key advantage of MR imaging guidance is that this technique provides excellent tissue contrast without contrast medium, which allows one to pass soft tissue structures that are often critical of nature (nerves vessels) [39,40].

Conversely, some limitations of MR imaging guidance must also be considered. The MR imaging surrounding sets specific requirements for procedures and the instrumentation and supporting hardware must be MR imaging safe. Local magnetic field inhomogeneities are induced resulting in position-encoding artifacts and in signal voids in the surrounding of instruments and especially near their tips. The artifacts generated by the susceptibility of the material are not only dependent on the material properties themselves and on the applied MR imaging sequences and parameters, but also on the geometric shape of the instruments and on the orientation to the static magnetic field in the MR imaging unit. When the latter is vertically oriented, needle angulation away from the vertical plane is crucial to maintain needle contrast [39,40]. Optical tracking for instrument guidance can be used because it facilitates real-time control of instrumentation position in relation to the image data [41,42]. Finally, MR imaging guidance seems more time-consuming than fluoroscopy or CT guidance even if the length of the procedural time decreases gradually as a learning-curve development in implementing the procedure.

Indications

The excellent tissue contrast provided by MR imaging explains the potential usefulness of MR imaging–guided biopsies of soft tissue lesions (especially inflammatory lesions or small tumoral lesions) and of bone lesions (especially when the trabecular bone destruction is not extensive enough to be detected correctly by the other types of imaging guidance). This technique has been applied to many other indications, however, including percutaneous bone biopsies [39,40], nerve root infiltration therapy [43, 44], arthrography, and discography [41]. The MR imaging guidance has also been applied successfully to percutaneous cryotreatment and interstitial laser treatment of osteoid osteoma [42]. The thermal monitoring capability of a low-field MR imaging scanner for radiofrequency ablation treatments has also been reported. Indeed, the use of some sequences including the completely balanced steady-state sequences may provide quantitative information about thermal changes in tissue, because the hyperintense region corresponds to the border of irreversible cell damage. Reliable signal change correlating to the change of temperature during treatment, however, has not been reported by all the studies [42].

Kinematic studies

Open-configured MR imaging scanners enable the achievement of kinematic studies. Besides the analysis of normal motion, this property has been especially reported at the patellofemoral joint to determine the presence and severity of patellar malalignment and abnormal tracking patterns [45], and at the lumbar spine for the acquisition of functional flexion-extension images [46]. Kinematic studies of the cervical spine in patients with rheumatoid arthritis may also be useful.

Summary

Recent emphasis on cost containment, the development of interventional techniques, the increased use of MR imaging for patients in intensive care and operating suites, the deployment of magnets in office suites, and the development of new magnet configurations all favor the supplementary use of low-field systems. Implementation of such systems may be useful, especially in orthopedic centers, or if installation of an additional high-field scanner is not possible because of economic considerations [47].

References

[1] Friedman DP, Rosetti GF, Flanders AE, et al. MR imaging: quality assessment method and ratings at 33 centers. Radiology 1995;196:219–26.

[2] Kinnunen J, Bondestam S, Kivioja A, et al. Diagnostic performance of low field MRI in acute knee injuries. Magn Reson Imaging 1994;12:1155–60.

[3] Cotten A, Delfaut E, Demondion X, et al. MR imaging of the knee at 0.2 and 1.5 T: correlation with surgery. AJR Am J Roentgenol 2000;174:1093–7.

[4] Parizel PM, Dijkstra HA, Geenen GP, et al. Low-field versus high-field MR imaging of the knee: a comparison of signal behaviour and diagnostic performance. Eur J Radiol 1995;19:132–8.

[5] Kladny B, Gluckert K, Swoboda B, et al. Comparison of low-field (0.2 Tesla) and high-field (1.5 Tesla) magnetic resonance imaging of the knee joint. Arch Orthop Trauma Surg 1995;114:281–6.

[6] Kersting-Sommerhoff B, Hof N, Lenz M, et al. MRI of peripheral joints with a low-field dedicated system: a reliable and cost-effective alternative to high-field units? Eur Radiol 1996;6:561–5.

[7] Sugimoto H, Hirose I, Miyaoka E, et al. Low-field-strength MR imaging of failed hip arthroplasty: association of femoral periprosthetic signal intensity with radiographic, surgical, and pathologic findings. Radiology 2003;229:718–23.

[8] Woertler K, Strothmann M, Tombach B, et al. Detection of articular cartilage lesions: experimental evaluation of low- and high-field-strength MR imaging at 0.18 and 1.0 T. J Magn Reson Imaging 2000;11: 678–85.

[9] Ma QY, Chan KC, Kacher DF, et al. Superconducting RF coils for clinical MR imaging at low field. Acad Radiol 2003;10:978–87.

[10] Huegli RW, Tirman PF, Bonel HM, et al. Use of the modified three-point Dixon technique in obtaining T1-weighted contrast-enhanced fat-saturated images on an open magnet. Eur Radiol 2004;14:1781–6.

[11] Bredella MA, Losasso C, Moelleken SC, et al. Three-point Dixon chemical-shift imaging for evaluating articular cartilage defects in the knee joint on a low-

field-strength open magnet. AJR Am J Roentgenol 2001;177:1371–5.

[12] Maas M, Dijkstra PF, Akkerman EM. Uniform fat suppression in hands and feet through the use of two-point Dixon chemical shift MR imaging. Radiology 1999;210:189–93.

[13] Wohlgemuth WA, Roemer FW, Bohndorf K. Short tau inversion recovery and three-point Dixon water-fat separation sequences in acute traumatic bone fractures at open 0.35 Tesla MRI. Skeletal Radiol 2002;31:343–8.

[14] Riel KA, Reinisch M, Kersting-Sommerhoff B, et al. 0.2-Tesla magnetic resonance imaging of internal lesions of knee joint : a prospective arthroscopically controlled clinical study. Knee Surg Sports Traumatol Arthrosc 1997;7:37–41.

[15] Kreitner KF, Hansen M, Schadmand-Fischer S, et al. Low-field MRI of the knee joint: results of a prospective, arthroscopically controlled study. Rofo Fortschr Geb Rontgenstr Neuen Bildgeb Verfahr 1999;170: 35–40.

[16] Rand T, Imhof H, Turetschek K, et al. Comparison of low field (0.2T) and high field (1.5T) MR imaging in the differentiation of torn from intact menisci. Eur J Radiol 1999;30:22–7.

[17] Ahn JM, Kwak SM, Kang HS, et al. Evaluation of patellar cartilage in cadavers with a low-field-strength extremity-only magnet: comparison of MR imaging sequences, with macroscopic findings as the standard. Radiology 1998;208:57–62.

[18] Harman M, Ipeksoy U, Dogan A, et al. MR arthrography in chondromalacia patellae diagnosis on a low-field open magnet system. Clin Imaging 2003;27:194–9.

[19] Kreitner KF, Loew R, Runkel M, et al. Low-field MR arthrography of the shoulder joint: technique, indications, and clinical results. Eur Radiol 2003;13: 320–9.

[20] Petersilge CA, Lewin JS, Duerk JL, et al. MR arthrography of the shoulder: rethinking traditional imaging procedures to meet the technical requirements of MR imaging guidance. AJR Am J Roentgenol 1997; 169:1453–7.

[21] Shellock FG, Bert JM, Fritts HM, et al. Evaluation of the rotator cuff and glenoid labrum using a 0.2-Tesla extremity magnetic resonance (MR) system: MR results compared to surgical findings. J Magn Reson Imaging 2001;14:763–70.

[22] Merl T, Scholz M, Gerhardt P, et al. Results of a prospective multicenter study for evaluation of the diagnostic quality of an open whole-body low-field MRI unit: a comparison with high-field MRI measured by the applicable gold standard. Eur J Radiol 1999; 30:43–53.

[23] Loew R, Kreitner KF, Runkel M, et al. MR arthrography of the shoulder: comparison of low-field (0.2 T) vs high-field (1.5 T) imaging. Eur Radiol 2000;10: 989–96.

[24] Allmann KH, Walter O, Laubenberger J, et al. Mag-

netic resonance diagnosis of the anterior labrum and capsule: effect of field strength on efficacy. Invest Radiol 1998;33:415–20.

[25] Tung GA, Entzian D, Green A, et al. High-field and low-field MR imaging of superior glenoid labral tears and associated tendon injuries. AJR Am J Roentgenol 2000;174:1107–14.

[26] Lepisto J, Mattila K, Nieminen S, et al. Low field MRI and scaphoid fracture. J Hand Surg [Br] 1995; 20:539–42.

[27] Raby N. Magnetic resonance imaging of suspected scaphoid fractures using a low field dedicated extremity MR system. Clin Radiol 2001;56:316–20.

[28] Herber S, Kreitner KF, Kalden P, et al. Low-field MRI of the ankle joint: initial experience in children and adolescents using an open 0.2 T MR-system. Rofo Fortschr Geb Rontgenstr Neuen Bildgeb Verfahr 2000; 172:267–73.

[29] Althoff CE, Hermann KG, Scheel AK, et al. Low-field magnetic resonance imaging for follow-up analysis of finger joint inflammation in patients with active rheumatoid arthritis receiving adalimumab [abstract]. Radiology 2004;303.

[30] Bonel HM, Schneider P, Seemann MD, et al. MR imaging of the wrist in rheumatoid arthritis using gadobenate dimeglumine. Skeletal Radiol 2001;30: 15–24.

[31] Palosaari K, Tervonen O. Post-processing water-fat imaging technique for fat suppression in a low-field MR imaging system, evaluation in patients with rheumatoid arthritis. MAGMA 2002;15:1–9.

[32] Yu W, Feng F, Dion E, et al. Comparison of radiography, computed tomography and magnetic resonance imaging in the detection of sacroiliitis accompanying ankylosing spondylitis. Skeletal Radiol 1998;27: 311–20.

[33] Bonel H, Helmberger T, Geiss HC, et al. Comparison of sequences for depicting bone marrow alterations in osteomyelitis applied in a low field strength magnetic resonance imaging system. MAGMA 1998;7: 1–8.

[34] Hovi I, Valtonen M, Korhola O, et al. Low-field MR imaging for the assessment of therapy response in musculoskeletal infections. Acta Radiol 1995;36:220–7.

[35] Eskelin MK, Lotjonen JM, Mantysaari MJ. Chronic exertional compartment syndrome: MR imaging at 0.1 T compared with tissue pressure measurement. Radiology 1998;206:305–7.

[36] Steinborn M, Heuck A, Jessel C, et al. Magnetic resonance imaging of lateral epicondylitis of the elbow with a 0.2-T dedicated system. Eur Radiol 1999; 9:1376–80.

[37] Hottya GA, Peterfy CG, Uffmann M, et al. Dedicated extremity MR imaging of the foot and ankle. Eur Radiol 2000;10:467–75.

[38] Ahn JM, Sartoris DJ, Kang HS, et al. Gamekeeper thumb: comparison of MR arthrography with conventional arthrography and MR imaging in cadavers. Radiology 1998;206:737–44.

[39] König CW, Duda SH, Trübenbach J, et al. MR-guided biopsy of musculoskeletal lesions in a low-field system. J Magn Reson Imaging 2001;13:761–8.

[40] König CW, Trübenbach J, Bohm P, et al. Magnetic resonance-guided transcortical biopsy of bone marrow lesions using a magnetic resonance imaging-compatible piezoelectric power drill: preliminary experience. Invest Radiol 2003;38:159–63.

[41] Sequeiros RB, Klemola R, Ojala R, et al. Percutaneous MR-guided discography in a low-field system using optical instrument tracking: a feasibility study. J Magn Reson Imaging 2003;17:214–9.

[42] Sequeiros RB, Hyvonen P, Sequeiros AB, et al. MR imaging-guided laser ablation of osteoid osteomas with use of optical instrument guidance at 0.23 T. Eur Radiol 2003;13:2309–14.

[43] Sequeiros RB, Ojala RO, Klemola R, et al. MRI-guided periradicular nerve root infiltration therapy in low-field (0.23-T) MRI system using optical instrument tracking. Eur Radiol 2002;12:1331–7.

[44] Ojala R, Klemola R, Karppinen J, et al. Sacro-iliac joint arthrography in low back pain: feasibility of MRI guidance. Eur J Radiol 2001;40:236–9.

[45] Shellock FG, Stone KR, Crues JV. Development and clinical application of kinematic MRI of the patello-femoral joint using an extremity MR system. Med Sci Sports Exerc 1999;31:788–91.

[46] Harvey SB, Smith FW, Hukins DW. Measurement of lumbar spine flexion-extension using a low-field open-magnet magnetic resonance scanner. Invest Radiol 1998;33:439–43.

[47] Wutke R, Fellner FA, Fellner C, et al. A low-field MR system in acute traumatological imaging in radiology: technical note. Rontgenpraxis 2001;54:43–8.

ELSEVIER
SAUNDERS

Radiol Clin N Am 43 (2005) 683 – 692

RADIOLOGIC
CLINICS
of North America

Shoulder MR Arthrography: How, Why, When

Marlena Jbara, MD, Qi Chen, MD, Paul Marten, MD[a], Morcos Morcos, MD[a],
Javier Beltran, MD[a,b],*

[a]Maimonides Medical Center, 4802 Tenth Avenue, Brooklyn, NY 11219, USA
[b]State University of New York Downstate, Brooklyn, NY, USA

The advent of advanced MR imaging sequence selection, the use of gadopentetate dimeglumine–based contrast agents, and an improved gradient strength for diagnosing pathology of the shoulder have provided the clinician with an armamentarium of diagnostic tools for accurate early diagnosis of capsulolabral pathology and disorders of the rotator cuff. With the increasing availability of direct and indirect MR arthrographic techniques, coupled with improvements in arthroscopic shoulder technique, the need has emerged to provide guidelines to stratify patients for conventional MR imaging and indirect or direct MR arthrography [1–4]. This article reviews current techniques for shoulder imaging, discusses advantages and disadvantages, and reviews the literature regarding sensitivity and specificity in the evaluation of shoulder pathology, specifically, glenoid labral tears and rotator cuff tears.

Conventional MR imaging of the shoulder

Optimization of the imaging technique depends on patient position and coil and sequence selection to obtain a superior signal-to-noise ratio. Sequence acquisition should be performed in the axial, sagittal, and coronal planes using the supraspinatus muscle and glenohumeral joint as anatomic landmarks to evaluate suspected pathology. As is true for all MR imaging, the sequence selection, field of view (FOV), slice thickness, and matrix are parameters that are manipulated to

give an optimal signal-to-noise ratio in a reasonable scan time. The ultimate goal is to have superior image quality in the shortest time possible to alleviate motion artifact and improve patient throughput.

Conventional MR imaging of the shoulder begins with the application of a phased array coil covering as much tissue as necessary. The smallest FOV is chosen to achieve the highest resolution, usually 14 to 16 cm. The slice thickness for shoulder sequences ranges from 2 to 3 mm, restricting the slice gap to as little as possible without compromising overall image quality. With the advent of high-resolution fast turbo spin-echo and proton density sequences, matrix pixels of 512×512 are routinely used without lengthening the scan time when compared with conventional spin-echo sequences. At the authors' center, nasal cannula oxygen is administered to encourage nasal breathing, which limits chest excursion, decreasing patient motion artifact.

From a coronal scout localizer, axial sections are obtained from the acromioclavicular (AC) joint through the glenoid margin, allowing a depiction of the capsular structures, subscapularis muscle, and long head of the biceps tendon as it courses through the bicipital groove. Proton density fat-suppressed axial images allow for improved detection of anterior and posterior labral pathology. Coronal oblique images are graphically prescribed from axial images parallel to the axis of the supraspinatus tendon for assessment of the rotator cuff, superior and inferior labrum, and AC joint and subjacent bursae. At most institutions, the coronal oblique sequence consists of a T2 fat-suppressed or short tau inversion recovery (STIR) protocol optimizing the conspicuity of pathology, because fluid and rotator cuff pathology is of

---

* Corresponding author. Maimonides Medical Center, 4802 Tenth Avenue, Brooklyn, NY 11219.
   E-mail address: jbeltran46@msn.com (J. Beltran).

higher signal intensity than the surrounding tissues. In additional, a T1-weighted coronal spin-echo is performed to evaluate for glenohumeral and AC joint arthrosis. Sagittal oblique images consist of proton density non–fat-suppressed and T2-weighted fat-suppressed images, which complement evaluation of the capsulolabral ligamentous complex, rotator cuff atrophy, and AC joint hypertrophy.

Indirect MR arthrography

Indirect MR arthrography takes advantage of bulk flow and the diffusion of contrast material from the vascular supply into the synovial tissue lining the bursae, joint capsule, and tendon sheaths, ultimately pooling into the joint space. Gadopentetate dimeglumine–based contrast agents shorten the T1 relaxation time of tissues, which can be used to produce arthrographic T1-weighted fat-suppressed images. This powerful technique enables an anatomic and physiologic assessment of joint pathology.

After sterilization of the injection site, usually the antecubital fossa, a dilute 15-mL solution of 0.1 mmol/kg gadopentetate dimeglumine mixed with saline is injected intravenously. Greater concentrations of gadopentetate dimeglumine including 0.2 and 0.4 mmol/kg have not been shown to derive greater arthrographic benefit [5]. Following 10 minutes of active exercise, axial, sagittal, and coronal T1-weighted fat-suppressed images, coronal T2 fat-suppressed fast spin-echo for the identification of extra-articular fluid collections, and abduction external rotation (ABER) T1-weighted fat-suppressed images are submitted for review.

Use of the ABER position in which the palm of the hand is positioned against the dorsal aspect of the craniocervical junction allows for a static assessment of stress placed on the anterior glenoid [6]. Triplane scout images of the patient in the ABER position are obtained from which sagittal images are prescribed, resulting in an oblique sagittal evaluation from the anteroinferior labrum through the posterosuperior labrum. Distraction of the joint as a result of this position allows visualization of partially healed labral-ligamentous lesions of the inferior glenoid rim that may otherwise be inconspicuous on conventional orthogonal sequences.

Direct MR arthrography

Direct MR arthrography is a two-phase procedure in which the intra-articular injection of contrast material is performed under fluoroscopic visualization, followed by transfer of the patient to the MR scanner for diagnostic imaging. The technique of intra-articular injection must avoid the cartilage, labrum, and capsular attachments to yield diagnostic utility [1,7–9]. Although multiple techniques have been described, the anterior approach is most common, which has been modified overtime [1,7].

The anterior approach to glenohumeral joint injection requires supine positioning of the patient with the shoulder in external rotation. External rotation exposes more of the articular surface of the humeral head anteriorly and increases the intra-articular area available for needle insertion. The first step requires localization of the desired needle position, which is medial to the superior third of the humeral head, which is covered by the joint capsule. The area is prepared and draped in a sterile fashion, and the subcutaneous tissue is anesthetized. Needle size may vary, but commonly a 20- to 22-gauge 3.5-inch spinal needle is used [1]. The needle tip is advanced in an anteroposterior direction to the humeral head, avoiding contact with the glenoid labrum.

In the posterior approach to shoulder arthrography, the patient is positioned prone with the ipsilateral shoulder raised off the table with a pad [1]. The needle is directed toward the inferomedial aspect of the humeral head. After local anesthetic is used, a 21-gauge spinal needle is advanced vertically under fluoroscopic guidance toward the cartilage of the humeral head [10].

Chung et al have demonstrated that the anterior approach may result in penetration of the anterior stabilizing structures of the glenohumeral joint, which has a tendency for anterior instability [7]. Their study included examination of six shoulders from fresh cadavers using an 18-gauge needle with a marker in the anterior and posterior approach. The marker for the anterior approach traversed through the subscapularis muscle or tendon in all cases, the inferior glenohumeral ligament in two cases, and the anterior inferior labrum. The marker for the posterior approach traversed the posteroinferior labrum in a single case without violation of the anterior structures. It has been recommended that one use the posterior approach if the patient presents with anterior instability or anterior symptoms [7,8]. The posterior approach will decrease the likelihood of injecting into the subscapularis tendon or inferior glenohumeral ligament [4].

Upon confirming the intra-articular location using the least iodinated contrast possible (2–3 mL), approximately 15 mL of a 0.1-mmol/kg solution of gadopentetate dimeglumine is injected, and the patient is taken to the MR scanner for sequence acquisition [10–13]. The volume of the injection ranges from 10 to 20 mL [1,12,14]. Injections of less than 15 mL decrease the likelihood of extra-articular leakage, which may be mistaken for a full-thickness rotator cuff tear [14]. Exercise after shoulder arthrography has no beneficial or detrimental effect on MR imaging quality or the depiction of a rotator cuff tear [15].

Following MR scout acquisition, axial, sagittal, and coronal T1-weighted fat-suppressed images, coronal T2 fat-suppressed fast spin-echo for the identification of extra-articular fluid collections, and ABER T1-weighted fat-suppressed images are submitted for review. Use of the ABER position during imaging improves the detection of tears or defects of the anterior glenoid labrum [6,16]. Additional maneuvers such as arm traction using 1- to 3-kg weights applied to the wrist combined with external rotation improve the sensitivity and specificity for detecting superior labral anteroposterior (SLAP) lesions [17].

Indirect versus direct shoulder arthrography: relative advantages and disadvantages of each method

Several factors beyond image quality must be considered when choosing indirect versus direct shoulder MR arthrography. These factors include the patient's adversity to the procedure; the safety, cost, extra time, and labor hours per procedure; and scheduling and coordinating the use of fluoroscopy and MR imaging and the direct involvement of the radiologist.

Adversity to the procedure

Postprocedure assessment of pain and discomfort show that direct arthrography is better tolerated than MR imaging. The patient's assessment of discomfort during direct MR arthrography of the shoulder using a visual analogue scale (VAS) was assessed by Binkert et al for 202 consecutive patients. The expected discomfort and that actually experienced during arthrography and MR imaging were compared. The postprocedure average VAS score was lower for the arthrographic procedure than for the MR imaging. The discomfort experienced during arthrography was worse than expected in 1% of patients (n = 2). Arthrography-related discomfort was generally well tolerated and often less severe than anticipated [17–19]. In opposition, indirect arthro-

graphy is minimally invasive with a peripheral intravenous injection of dilute gadopentetate dimeglumine. In the authors' experience, this procedure is well tolerated, with no significant complications elicited.

Safety

Direct arthrography exposes the patient to ionizing radiation. Usually, the procedures are straightforward and do not require much time under fluoroscopy; however, if complications arise, the amount of time and radiation dose (3 rad/min) from fluoroscopy can be considerable. Additionally, the potential for a contrast reaction is considerably greater for the iodinated agents. The danger and complication of a short-term peripheral intravenous injection is relatively negligible during indirect arthrography. Although the overall risk of complications with minimally invasive intra-articular needle arthrography is low, the potential complications are considerable. An introduced infection to the joint can result in septic arthritis, osteomyelitis, or fasciitis. Arthrography may cause direct damage to the nerves, capsule, or ligaments.

Cost and simplicity

Indirect arthrography is less costly per procedure, requiring less MR scan time and avoiding the cost of iodinated contrast. For direct arthrography, the expense of the fluoroscopy suite, nursing, and MR imaging scan time decrease the efficiency when evaluating patient throughput. The cost of the radiologist's time for direct arthrography is considerably greater than when a nurse or technologist is used to deploy the intravenous injection.

Utility

Indirect MR arthrography allows an assessment of intra-articular and extra-articular soft tissues (Fig. 1) [20,21]. It is based on the concept that the synovial membrane is vascular, and that injected intravenous contrast will diffuse to the joint overtime. This property is advantageous when diagnosing inflammatory arthropathies such as rheumatoid arthritis, which result in synovial hyperplasia. The contrast enhances this intermediate-signal synovial tissue, a finding that is conspicuous during indirect arthrographic imaging. A potential drawback of indirect arthrography is the inability to control the volume of the contrast that diffuses into the joint. Tense effusions with increased viscosity (hemorrhagic fluid) will decrease contrast diffusion into the joint. Nev-

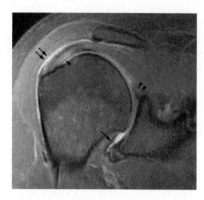

Fig. 1. Normal indirect MR arthrogram. Oblique coronal T1-weighted fat-saturated image following intravenous administration of gadolinium demonstrates small amount of contrast in the axillary recess (*single arrow*), enhancement of the subacromial subdeltoid bursa (*double arrow*), and magic angle artifact in the distal supraspinatus tendon (*single arrowhead*). The superior labrum is somewhat indistinct and bright (*double arrowhead*) owing to magic angle artifact.

ertheless, this may be advantageous as joint diffusion is increased with vascularized inflamed joints or after gentle exercise because of the hyperemia. Indirect arthrography gives more data regarding the physiologic state as well as assessing intra-articular surface anatomy [13]. Indirect MR arthrography does not require fluoroscopic guidance or joint injection, and it is superior to conventional MR imaging in delineating the labrum when there is minimal joint fluid [14,21].

The primary advantage of direct MR arthrography is its ability to detect capsulolabral pathology and partial-thickness and vertical rotator cuff tendon tears. Better distention of the joint capsule, particularly the labral-ligamentous complex, allows easier depiction of irregular tears versus more smoothly delineated anatomic variants such as sublabral sulci and foramina. Ideally, MR imaging should begin within 30 minutes of joint injection to minimize the absorption of contrast. This requirement makes the timing of direct arthrography more crucial and difficult to coordinate in a busy hospital setting. In comparison, the lack of controlled capsular distension and less reliable outlining of articular structures without exercise places indirect arthrography at a disadvantage.

Direct arthrography can cause considerable artifact. Leakage of contrast through the capsular puncture site can cause spread of contrast along the fascial planes into the subdeltoid space, causing a "bursogram," which can be misinterpreted as a full-thickness rotator cuff tear. Accidental injection of gas

can lead to an incorrect diagnosis of loose bodies from the magnetic susceptibility artifact. Attributing the exact cause of a susceptibility artifact should be based on its location; joint bodies are typically located in the more dependent portions, whereas gas bubbles rise to the nondependent portions of the joint. There are also some limitations in the detection of rotator cuff pathology using indirect arthrography. For example, the enhancement of subacromial bursa can obscure a rotator cuff tear. This potential pitfall may be avoided by comparing pre- and postcontrast images [21].

Direct versus indirect MR arthrography: literature review

In recent years, numerous articles have been published evaluating the sensitivity, specificity, and accuracy of nonenhanced indirect and direct MR arthrography in the evaluation of tears of the glenoid labrum and rotator cuff. Most of these articles have shown superior sensitivity and accuracy of direct MR arthrography over conventional MR imaging, particularly for SLAP lesions and partial tears of the rotator cuff muscles [22–31]. Nevertheless, as improvements in software design and surface coils continue, and as the common imaging interpretation pitfalls are better understood, the relative accuracy of each MR approach may change. The authors performed a PubMed search for "MRI arthrography shoulder" from January 1999 to November 2004. Articles that dealt with nonenhanced MR imaging and direct or indirect MR arthrography were reviewed and compared with each other or with arthroscopic or open surgery.

Labral tears

Table 1 compares recent articles that attempted to identify glenoid labral lesions using direct, indirect, or nonenhanced MR imaging. All of the included articles had an arthroscopic correlation or open surgery correlation. All were preoperative diagnoses except for one. Three articles compared indirect MR arthrography with nonenhanced MR imaging [32–35]. Another article compared direct MR arthrography with arthroscopic or open surgery [36–38]. One article compared nonenhanced MR imaging with the findings on arthroscopic surgery [39]. Another article compared all three MR approaches in an attempt to identify recurrent labral tears before second-look surgery [40]. A recent article compared direct MR arthrography with nonenhanced

Table 1
Comparative sensitivity, specificity and accuracy using nonenhanced, indirect, and direct MR arthrography in the evaluation of glenoid-labrum lesions

Study parameters	Labral tear	Labral lesion	SLAP	SLAP	SLAP	SLAP	Superior labral lesion	Recurrent labral tear
Prospective (no. of patients)	28	52	35	—	52	—	102	—
Retrospective				80		28		24
Direct MR arthrography (% sensitivity/ specificity/accuracy)	—	—	—	92/82/85	89/91/90	50/86/79	—	100/60/67
Indirect MR arthrography	90, 89	100, 71	91,85,89	—	—	—	—	100,100,100
Unenhanced MR	79, 67	70, 71	73,85,77	—	—	—	98,90,96	71, 80, 75
Arthroscopic correlation (no. of patients)	Y	Y	Y	80	Y	28	102	24
Surgical correlation	Y	Y	—	—	Y	—	—	—
Study	Sommer et al [32]	Maurer et al [35]	Herold et al [34]	Jee et al [36]	Bencardino et al [38]	Lee et al [37]	Connell et al [39]	Wagner et al [40]

Abbreviation: Y, presence of correlation but nonspecified number.

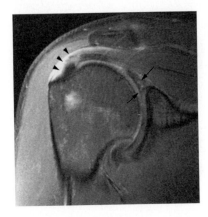

Fig. 2. SLAP lesion and partial rotator cuff tear demonstrated with indirect MR arthrogram. Oblique coronal T1-weighted fat-saturated image following intravenous administration of gadolinium demonstrates well-defined linear enhancement within the superior labrum, representing a SLAP lesion (*arrows*). There is irregularity of the bursal surface of the supraspinatus tendon (*arrowheads*), representing a partial bursal side tear.

MR imaging in two population groups—professional baseball players and nonprofessional athletes [41].

The overall sensitivity of direct arthrography has not changed in these articles when compared with the 91% sensitivity reported by Palmer et al in 1994 [22]. The low sensitivity reported in the article by Lee et al [37] is atypical and could have occurred owing to several reasons, including the fact that no exercise was performed before obtaining the arthrographic images, and provocative maneuvers such as arm traction were not used to make tears more conspicuous. Lee et al [37] suggested that if technical and perceptual error had been corrected, the sensitivity could have been around 83%.

The four articles that included indirect MR arthrography reported excellent sensitivity ranging from 90% to 100%. The specificity was not as good as that with direct arthrography, ranging from 71% to 89%. Based on a small sample size of six patients, Wagner et al [40] reported a 100% sensitivity using indirect MR arthrography postoperatively.

Five of the nine articles were prospective studies. Only one of these articles compared direct MR arthrography with indirect MR arthrography head-to-head in the same study [40]. This article found indirect MR arthrography to be an accurate means (100% sensitivity and 100% specificity) of evaluating the shoulder following instability surgery. Direct arthrography was reported to have 100% sensitivity but only 60% specificity. Wagner et al [40] explained that this lower specificity was a consequence

of two false-positive cases in a small sample size (6 patients). These two false-positive cases were due to metallic artifact from an anterior glenoid anchor following Bankart repair (mimicking a recurrent anterior labral tear) and a frayed and irregular labrum being overread.

In four of the five articles that evaluated nonenhanced MR imaging for labral tears, sensitivity ranged from 70% to 79% and specificity from 67% to 85%. The statistics were similar in diagnosing preoperative and postoperative labral tears. Nevertheless, in a prospective study involving a relatively large population size of 104 patients, Connell et al [39] reported sensitivity of 98%, specificity of 90%, and accuracy of 96% using nonenhanced MR. The high accuracy obtained in this study may have been due to several technical improvements, such as the use of phased array surface coils to improve the signal-to-noise ratio, and increasing the matrix number, leading to a smaller pixel size and, consequently, better spatial resolution (Figs. 2 and 3).

In a recent retrospective study, Magee et al [41] compared direct MR arthrography with nonenhanced MR imaging of the shoulder in 20 consecutive professional baseball players with shoulder pain. These findings were compared with those in a control group of 50 consecutive nonprofessional athletes with shoulder pain. Eleven of the fourteen professional athletes who had additional findings on direct MR arthrography but not on nonenhanced MR imaging had an arthroscopic correlation. All five of the nonprofessional athletes with additional findings on direct MR arthrography had an arthroscopic correla-

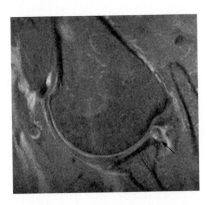

Fig. 3. SLAP lesion demonstrated in the ABER position with indirect MR arthrogram. ABER T1-weighted fat-saturated image following intravenous administration of gadolinium demonstrates a posterosuperior labral tear (*arrow*). Note the minimal amount of contrast in the joint and minimal enhancement of the surrounding soft tissues that are sufficient to demonstrate the SLAP lesion.

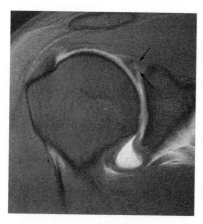

Fig. 4. SLAP lesion demonstrated with direct MR arthrogram. Oblique coronal T1-weighted fat-saturated image following intravenous administration of gadolinium. Arrows indicate abnormal signal intensity in the superior labrum corresponding to a SLAP lesion (type 1). Note the similar appearance between this direct MR arthrogram and the indirect MR arthrogram demonstrated in Fig. 2. Note also the extravasation of contrast material along the subscapularis muscle area.

tion. In the group of professional athletes, nonenhanced MR imaging missed six of the eight SLAP lesions, two of the three anterior labral lesions, and one of the two posterior labral lesions picked up by direct MR arthrography. Nine additional labral lesions were picked up by direct MR arthrography but not by nonenhanced MR imaging. This difference translates into a higher sensitivity of three folds for SLAP lesions, two folds for anterior labral lesions, and one fold for posterior labral lesions when using direct MR arthrography versus nonenhanced MR imaging in professional athletes. The relative yield of direct MR arthrography versus nonenhanced MR

imaging of the shoulder is less pronounced in nonprofessional athletes. In this population group, Magee et al [40] found that nonenhanced MR imaging missed two of five anterior labral tears and two of ten SLAP tears that were present on direct MR arthrography. Direct MR arthrography may be more effective in high-performance athletes for several reasons. Smaller tears, a greater number and higher frequency of labral tears, and supraspinatus tears often occur in atypical locations in these individuals (Fig. 4) [34,35,42–45].

Rotator cuff tears

Table 2 compares recent articles that have looked at direct, indirect, or nonenhanced MR imaging for the evaluation of rotator cuff tears. To the authors' knowledge, there has been no article during the last 5 years on the preoperative diagnosis of rotator cuff tears using direct MR arthrography. In 1994, Palmer et al reported a specificity of 75% to 100% in the preoperative diagnosis of rotator cuff tears using direct arthrography [23]. Wagner et al reported similar specificity postoperatively, but sensitivity data were not obtainable, because all cases in the direct MR arthrography group were true negatives [40]. In a prospective series of 30 patients after suture-anchor Bankart repair, Sugimoto et al [46] reported 100% sensitivity, 82% specificity, and 95% accuracy.

Two prospective studies evaluating indirect MR arthrography in the preoperative diagnosis of rotator cuff tears yielded sensitivities of 75% and 100%, with lower specificities of 50% and 83%, respectively [33,42]. A small retrospective study in postoperative patients demonstrated sensitivity of 67% and specificity of 100% [40].

Table 2
Comparative sensitivity, specificity, and accuracy using nonenhanced, indirect, and direct MR arthrography in the evaluation of rotator cuff lesions

Study parameters	Rotator cuff lesion	Rotator cuff tear	Recurrent rotator cuff tear	Intact suture
Prospective (no. of patients)	63	24	—	30
Retrospective	—	—	24	—
Direct MR arthrography (% sensitivity/specificity/accuracy)	—	—	NA/100/100	100/82/95
Indirect MR arthrography	75, 50	100, 83	67, 100, 83	—
Unenhanced MR	—	77, 83	33, 100, 83	—
Arthroscopic correlation (no. of patients)	—	—	—	30
Surgical correlation	32	24	24	—
Study	Rudolph et al [33]	Yagei et al [42]	Wagner et al [40]	Sugimoto et al [46]

Abbreviation: NA, not applicable.

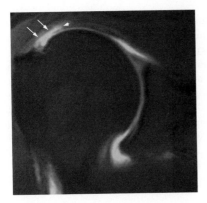

Fig. 5. Partial rotator cuff tear demonstrated with direct MR arthrogram. Oblique coronal T1-weighted fat-saturated image following intra-articular administration of gadolinium. Note the partial articular surface tear of the infraspinatus tendon (*arrows*) with delamination (*arrowhead*).

Nonenhanced MR imaging was evaluated preoperatively by Yagei et al [42] and postoperatively by Wagner et al [40]. The preoperative sensitivity was 77% and the specificity 83%. The postoperative sensitivity was 33% and the specificity 100%. The latter study identified only one of three rotator cuff tears [40].

In the evaluation of 20 professional athletes, Magee et al reported two full-thickness and six partial-thickness undersurface supraspinatus tendon tears missed by nonenhanced MR imaging but present on direct MR arthrography [41]. In this same study, among the 50 patients in the nonprofessional athlete group, two of six partial-thickness supraspinatus tendon tears were missed by nonenhanced MR imaging. Three full-thickness tears in this group were picked up by both methods (Fig. 5).

Summary

MR imaging is a powerful tool in the diagnosis and follow-up of patients with various shoulder pathologies. Attention to anatomic detail is paramount in obtaining diagnostic accuracy. Age and symptomatology should guide the choice of conventional MR imaging, indirect MR arthrography, or direct MR arthrography. The advantages and disadvantages of each technique have been discussed herein. Tables 1 and 2 compare recent publications that have used direct, indirect, or nonenhanced MR imaging in an attempt to identify glenoid labral lesions or rotator cuff tears, respectively. Nonenhanced MR imaging is much simpler to perform;

with increasing acuity as to normal variants and pitfalls, it remains the mainstay in diagnosing rotator cuff pathology. Its sensitivity in identifying capsulolabral lesions may be limited, particularly in the absence of native joint effusion. The accuracy of indirect MR arthrography is similar to that of direct MR arthrography except for one population group. Box 1 lists the various indications for each of the three MR techniques. Based on a meta-analysis, the authors believe that direct MR arthrography should be reserved for young athletes (less than 40 years old) who present with instability or chronic injuries. This group of patients seems to benefit most from direct MR arthrography because they tend to have smaller labral tears in greater number and frequency. Additionally, the accompanying rotator cuff pathology tends to occur in atypical locations [34,35,42–45]. Nonenhanced MR imaging may be sufficiently accurate in patients aged more than 40 years who have trauma or instability, in AC joint evaluation, and in other causes of shoulder pain, such as neoplasm, cervical spine disease, vascular disease, and

---

Box 1. Clinical scenarios and their recommended method of MR evaluation

Direct MR arthrography

> Athletes with chronic injuries (younger than 40 years)
> Instability (younger than 40 years)

Indirect MR arthrography

> Labral lesions, variant versus pathologic
> Paralabral cysts
> Biceps anchor
> Postoperative evaluation of labrum or rotator cuff
> Partial rotator cuff tear
> Chronic recurrent instability
> Inflammatory arthropathy

Nonenhanced MR

> Trauma (older than 40 years)
> Instability (older than 40 years)
> AC joint evaluation
> Other causes of shoulder pain: neoplasm, cervical spine disease, vascular disease, neural disease

neural disease. Indirect MR arthrography has a high diagnostic accuracy in the evaluation of pathologic versus normal anatomic variant labral lesions, paralabral cysts, and biceps anchors; in the postoperative evaluation of the labrum or rotator cuff; and the evaluation of partial rotator cuff tears, chronic recurrent instability, and inflammatory arthropathy.

References

[1] Jacobson JA, Lin J, Jamadar D, et al. Aids to successful shoulder arthrography performed with a fluoroscopically guided anterior approach. Radiographics 2003;23:373–9.

[2] Tirman PFJ, Steinbach LS, Belzer JP, et al. A practical approach to imaging of the shoulder with emphasis on MR imaging. Orthop Clin North Am 1997; 28(4):483–515.

[3] Stoller DW. MR arthrography of the glenohumeral joint. Radiol Clin North Am 1997;35:97–116.

[4] Matsuzaki S, Yoneda M, Kobayashi Y, et al. Dynamic enhanced MRI of the subacromial bursa: correlation with arthroscopic and histological findings. Skeletal Radiol 2003;32:510–20.

[5] Vahlensieck M, Peterfly CG, Wischer T, et al. Indirect MR arthrography: optimization and clinical applications. Radiology 1996;200:249–54.

[6] Lee SY, Lee JK. Horizontal component of partial-thickness tears of rotator cuff: imaging characteristics and comparison of ABER view with oblique coronal view at MR arthrography—initial results. Radiology 2002;224:470–6.

[7] Chung CB, Dwek JR, Feng S, et al. MR arthrography of the glenohumeral joint: a tailored approach. AJR Am J Roentgenol 2001;177:217–9.

[8] Helgason JW, Chandanani VP, Yu JS. MR arthrography: a review of current technique and applications. AJR Am J Roentgenol 1997;168:1473–9.

[9] Hajek PC, Baker LL, Sartoris DJ, et al. MR arthrography: anatomic-pathologic investigation. Radiology 1987;163:141–7.

[10] Kopka L, Funke M, Fischer U, et al. MR arthrography of the shoulder with gadopentetate dimeglumine: influence of concentration, iodinated contrast material, and time on signal intensity. AJR Am J Roentgenol 1994;163:621–3.

[11] Binkert CA, Zanetti M, Hoder J. MR arthrography of the glenohumeral joint: two concentrations of gadoteridol versus Ringer solution as the intra-articular contrast material. Radiology 2001;220:219–24.

[12] Hajek PC, Sartoris DJ, Neumann CH, et al. Potential contrast agents for MR arthrography: in vitro evaluation and practical observations. AJR Am J Roentgenol 1987;149:97–104.

[13] Weinmann HJ, Brasch RC, Press WE, et al. Characteristics of gadolinium-DTPA complex; a potential NMR contrast agent. AJR Am J Roentgenol 1984;142: 619–24.

[14] Steinbach LS, Palmer WE, Schweitzer ME. Special focus session: MR arthrography. Radiographics 2002; 22:1223–46.

[15] Brenner ML, Morrison WB, Carrino JA, et al. Direct MR arthrography of the shoulder: is exercise prior to imaging beneficial or detrimental? Radiology 2000; 215:491–6.

[16] Choi JA, Suh SI, Kim BH, et al. Comparison between conventional MR arthrography and abduction and external rotation MR arthrography in revealing tears of the antero-inferior glenoid labrum. Korean Journal of Radiology 2001;2(4):216–21.

[17] Chan KK, Muldoon KA, Yeh L, et al. Superior labral anteroposterior lesions: MR arthrography with arm traction. AJR Am J Roentgenol 1999;173:1117–22.

[18] Hall FM, Goldberg RP, Wyshak G, et al. Shoulder arthrography: comparison of morbidity after use of various contrast media. Radiology 1985;154:339–41.

[19] Binkert CA, Zanetti M, Holder J. Patient's assessment of discomfort during MR arthrography of the shoulder. Radiology 2001;221:775–8.

[20] Farmer KD, Hughes PM. MR arthrography of the shoulder: fluoroscopically guided technique using a posterior approach. AJR Am J Roentgenol 2002;178: 433–4.

[21] Bergin D, Schweitzer ME. Indirect magnetic resonance arthrography. Skeletal Radiol 2003;32:551–8.

[22] Palmer WE, Brown JH, Rosenthal DI. Labral-ligamentous complex of the shoulder: evaluation with MR arthrography. Radiology 1994;190:645–51.

[23] Palmer WE, Brown JH, Rosenthal DI. Rotator cuff: evaluation with fat-suppressed MR arthrography. Radiology 1993;188:683–7.

[24] Flannigan B, Kursunoglu-Brahme S, Snyder S, et al. MR arthrography of the shoulder: comparison with conventional MR imaging. AJR Am J Roentgenol 1990;155:829–32.

[25] Hodler J, Kursunoglu-Brahme S, Snyder SJ, et al. Rotator cuff disease: assessment with MR arthrography versus standard MR imaging in 36 patients with arthroscopic confirmation. Radiology 1992;182: 431–6.

[26] Burk Jr DL, Karasick D, Mitchell DG, et al. Rotator cuff tears: prospective comparison of MR imaging with arthrography, sonography, and surgery. AJR Am J Roentgenol 1989;152:87–92.

[27] Farley TE, Neumann CH, Steinbach LS, et al. Full-thickness tears of the rotator cuff of the shoulder: diagnosis with MR imaging. AJR Am J Roentgenol 1992;158:347–51.

[28] Kaplan PA, Bryans KC, Davick JP, et al. MR imaging of the normal shoulder: variants and pitfalls. Radiology 1992;184:519–24.

[29] Robertson PL, Schweitzer ME, Mitchell DG, et al. Rotator cuff disorders: interobserver and intraobserver variation in diagnosis with MR imaging. Radiology 1995;194:831–5.

[30] Tsai JC, Zlatkin MB. Magnetic resonance imaging of the shoulder. Radiol Clin North Am 1990;28:279–91.

[31] Shankman S, Bencardino J, Beltran J. Glenohumeral instability: evaluation using MR arthrography of the shoulder. Skeletal Radiol 1999;28:365–82.

[32] Sommer T, Vahlensieck M, Wallny T, et al. Indirekte MR-arthrographie in der Diagnostik von Lasionen des Labrum glenoidale [Indirect MR arthrography in the evaluation of tears of the glenoid labrum]. Fortschr Rontgenstr 1997;167(1):46–51 [in German].

[33] Rudolph J, Lorenz M, Schroder R, et al. Indirekte MR-arthrographie in der diagnostik von rotatoren-manschettenlasionen [Detection of rotator cuff lesions with indirect MR arthrography]. Fortschr Rontgenstr 2000;172:686–91.

[34] Herold T, Hente R, Zorger N, et al. Indirekte MR-arthrographie der schulter-wertigkeit im nachweis von SLAP-lasionen [Indirect MR arthrography of the shoulder—value in the detection of SLAP lesions]. Fortschr Rontgenstr 2003;175:1508–14.

[35] Maurer J, Rudolph J, Lorenz M, et al. Prospektive studie zum nachweis von lasionen des labrum glenoidale mit der indirekten MR-arthrographie der schulter [A prospective study on the detection of labral lesions by indirect MR arthrography of the shoulder]. Fortschr Rontgenstr 1999;171:307–12.

[36] Jee WH, McCauley TR, Katz LD, et al. Superior labral anterior posterior (SLAP) lesions of the glenoid labrum: reliability and accuracy of MR arthrography for diagnosis. Radiology 2001;218:127–32.

[37] Lee JHE, Van Raalte V, Malian V. Diagnosis of SLAP lesions with Grashey-view arthrography. Skeletal Radiol 2003;32:388–95.

[38] Bencardino JT, Beltran J, Rosenberg ZS, et al. Superior labrum anterior-posterior lesions: diagnosis with MR arthrography of the shoulder. Radiology 2002;214:267–71.

[39] Connell DA, Potter HG, Wickiewicz TL, et al. Non-contrast magnetic resonance imaging of superior labral lesions: 102 cases confirmed at arthroscopic surgery. Am J Sports Med 1999;27(2):208–13.

[40] Wagner SC, Schweitzer ME, Morrison WB, et al. Shoulder instability: accuracy of MR imaging performed after surgery in depicting recurrent injury—initial findings. Radiology 2002;222:196–203.

[41] Magee T, Williams D, Mani N. Shoulder MR arthrography: which patient group benefits most? AJR Am J Roentgenol 2004;183:969–74.

[42] Yagei B, Manisah M, Yilmaz E, et al. Indirect MR arthrography of the shoulder in detection of rotator cuff ruptures. Eur Radiol 2001;11:258–62.

[43] Roger B, Skaf A, Hooper AW, et al. Imaging findings in the dominant shoulder of throwing athletes: comparison of radiography, arthrography, CT arthrography, and MR arthrography with arthroscopic correlation. AJR Am J Roentgenol 1999;172:1371–80.

[44] Tuite MJ. MR imaging of sports injuries to the rotator cuff. Magn Reson Imaging Clin N Am 2003;11:207–19.

[45] Farber JM, Buckwalter KA. Sports-related injuries of the shoulder: instability. Radiol Clin North Am 2002;40:235–49.

[46] Sugimoto H, Suzuki K, Mihara KI, et al. MR arthrography of shoulders after suture-anchor Bankart repair. Radiology 2002;224:105–11.

ELSEVIER
SAUNDERS

Radiol Clin N Am 43 (2005) 693 – 707

RADIOLOGIC
CLINICS
of North America

Ankle MR Arthrography: How, Why, When

Luis Cerezal, MD*, Faustino Abascal, MD, Roberto García-Valtuille, MD,
Ana Canga, MD

Department of Radiology, Instituto Radiológico Cántabro, Clínica Mompía, Mompía, 39109 Cantabria, Spain

MR imaging has become established as the most effective imaging method in the assessment of numerous disorders of the ankle joint. MR arthrography extends the capabilities of conventional MR imaging because intra-articular injection of contrast solution allows selective examination of a joint, with controlled capsular distention and excellent depiction of the internal structures. There are some features of MR arthrography, however, that limit its clinical use, including the conversion of a noninvasive procedure into a mildly invasive one, exposing patients to ionizing radiation and the risks of intra-articular needle placement, and the increased cost and time required to perform MR arthrography compared with conventional MR imaging. Despite these limitations, MR arthrography is being used increasingly to evaluate intra-articular pathology of the ankle.

Indirect MR arthrography with intravenous administration of gadolinium also leads to an enhancement effect of the joint cavity, but it lacks capsular distention. This MR imaging technique is considered an alternative to direct MR arthrography in some cases.

This article reviews the role of MR arthrography in the evaluation of the ankle joint, considering techniques, pitfalls, complications, pertinent anatomy, and applications. It also provides a brief overview of the usefulness of indirect MR arthrography in this joint.

How to perform ankle MR arthrography

MR arthrography of the ankle is a two-step procedure involving intra-articular injection of contrast solution before MR imaging. The injection can be performed in two main sites (Fig. 1A) at the anterior aspect of the ankle: immediately medial to the tibialis anterior tendon or medially to the tendon of the extensor hallucis longus [1–3]. The arthrogram usually is performed under fluoroscopy control; however, ultrasound, CT, or MR guidance also may be used [4–7].

The patient is placed in lateral decubitus position with the ankle in the lateral position and the front of the ankle facing the examiner. The course of the dorsalis pedis artery is palpated and marked to avoid its puncture. Using fluoroscopic guidance, a 23-gauge needle is inserted under sterile conditions into the tibiotalar joint medially to the tibialis anterior tendon with a slight cranial tilt to avoid the overhanging anterior margin of the tibia (Fig. 1B). Before the injection of contrast material, any fluid within the joint is aspirated so as not to dilute the contrast material. Intra-articular needle placement is confirmed with an injection of a drop of 2 mL of iodinated contrast material. If the needle is intra-articular, the contrast medium flows away from the needle tip and draws capsular recesses. Subsequently a mixture of 0.1 mL of gadolinium, 10 mL of saline solution, 5 mL of iodinated contrast material, and 5 mL of lidocaine 1% is injected until the joint capsule is properly distended (~10 mL). The presence of iodinated contrast material in the mixture ensures correct needle position and adequate capsular distention [8]. To prevent capsular disruption, contrast

---

* Corresponding author.
E-mail address: lcerezal@mundivia.es (L. Cerezal).

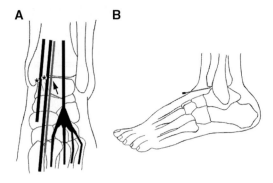

Fig. 1. Diagrams illustrate the injection sites for ankle joint MR arthrography. (*A*) Medially to anterior tibial tendon and medially to extensor hallucis longus tendon (*asterisk*). The course of the dorsalis pedis artery (*arrow*) should be avoided. (*B*) The needle is directed slightly cranially so that it can slide easily beneath the anterior lip of the tibia and advanced until its tip is seen between the distal tibia and the talus.

injection is stopped if the patient expresses discomfort or if high resistance is felt during the instillation of the solution. In a normal ankle, the injected contrast material forms an umbrella over the articular surface of the talus with prominent anterior and posterior capsular recesses. An upward extension of contrast material is seen between the distal tibia and fibula, filling the syndesmotic recess. In 25% of cases, contrast solution may enter the flexor hallucis longus (FHL) and flexor digitorum longus tendon sheaths and the subtalar joint [3,9,10]. There should be no tendon sheath filling on the lateral side of a normal ankle. After the injection, the needle is removed and the ankle is manipulated briefly to distribute the contrast medium uniformly.

Joint puncture can be performed in the MR imaging suite without radiologic guidance, using recognized anatomic landmarks and avoiding the need for iodinated contrast agents and ionizing radiation. It is particularly useful when there is limited access to a fluoroscopic suite [2]. The puncture point is located at the level of the anteromedial ankle joint, just medial to the tibialis anterior tendon, and about 5 mm proximal to the medial malleolus. After a brief learning period of fluoroscopic injection of the ankle joint, the injection can be performed easily without guidance with imaging methods by identifying anatomic landmarks.

Saline solution may be injected as the MR arthrographic contrast material. Normal saline is not an ideal contrast medium, however, because it has the same signal characteristics as preexisting joint effusion and para-articular fluid, and it is not possible to determine if they have occurred as a result of the saline injection [4–7].

Different studies have shown that patients who have undergone MR arthrography considered the discomfort less than expected [11,12]. No significant side effects have been reported that are attributable to intra-articular gadolinium solution [8,12]. The main complications of MR arthrography are joint pain that may last 1 to 3 days after joint puncture and vasovagal reactions. Articular distention in arthrography produces a feeling of pressure in the joint and pain on moving with variable intensity and progressive decrease in the first days after injection [12]. Vasovagal reactions may occur in patients undergoing arthrography, particularly in young athletic patients with low resting heart rates. Vasovagal reactions may be a result of coexisting circumstances, such as emotion, apprehension, and pain. Most patients who experience a vasovagal event have a quick recovery and are managed easily in the radiology suite. The routine administration of prophylactic atropine before ankle arthrography to block vasovagal reactions is unnecessary given the low incidence of these reactions (about 1% in the authors' experience). The number of vasovagal reactions diminishes considerably if the patient is not allowed to see the material of puncture and the procedure.

Joint infection is the major complication of arthrography; this extremely rare complication is independent of the type of substance being injected into the joint [12]. No reports exist on serious adverse events, such as anaphylactic shock or other events requiring treatment in an intensive care unit or hospitalization [12].

MR imaging should be performed shortly after intra-articular injection of contrast solution to minimize absorption of contrast and guarantee the desired capsular distention, although imaging delays of 2 hours are tolerated [3,7,12]. Imaging protocol should include axial, sagittal, and coronal planes with a dedicated extremity coil. Field of view may be decreased to optimize the visualization of intra-articular structures. T1-weighted spin-echo sequences with and without fat suppression maximize the signal intensity of contrast solution. Fat suppression is crucial in MR arthrography because fat and gadolinium have similar signal intensities on T1-weighted images, creating diagnostic difficulty. Fat suppression selectively decreases the signal from fat, while preserving the signal from contrast solution, confirming or excluding extra-articular contrast material. At least one sequence should be a fat-suppressed T2-weighted image for the detection of subtle bone marrow edema and extra-articular fluid collections

[2,3,5,7]. Several authors have postulated the use of oblique planes or forced positions of the foot in the assessment of the anterior talofibular (ATF) and calcaneofibular (CF) ligaments that follow an oblique course. The ATF ligament optimally is imaged with the patient's foot in plantar flexion on axial images or with the use of oblique axial images. The CF ligament optimally is imaged with the foot in dorsiflexion on axial images or with the use of oblique coronal images [13–16]. Three-dimensional gradient echo images allow reconstructions adapted to the anatomic course of these ligaments with good accuracy and render oblique planes or forced positions unnecessary. Three-dimensional gradient echo images also are helpful in the assessment of cartilage lesions and loose bodies. Overall, the choice of sequence depends on radiologist preference and sequence availability.

The most common pitfalls of MR arthrography of the ankle are extra-articular injection or leak of contrast material through the capsular puncture site, which can be confused with capsular disruption. Accumulation of contrast material in the anterior and posterior recesses of the tibiotalar joint, which manifests as smooth, encapsulated fluid outside the ligaments, can be interpreted as a ligamentous tear. The bulbous appearance of the posterior talofibular (PTF) ligament and posterior tibiofibular ligament on sagittal images can simulate loose bodies. This pitfall is avoided easily by the evaluation of consecutive sagittal images and knowledge of the location of these ligaments [17,18]. The pseudodefect of the talar dome is a normal groove at the posterior aspect of the talus. This defect should not be misinterpreted as an articular erosion or osteochondral defect [10,17,18]. The inadvertent use of undiluted gadolinium results in marked T1 and T2 shortening with fluid appearing low in signal. The instillation of air bubbles during injection may mimic loose bodies, although generally air bubbles rise to nondependent regions of the joint, whereas loose bodies gravitate to dependent locations.

Why and when to perform ankle MR arthrography

Ligamentous injuries

The ankle joint is supported by three ligamentous groups: the distal tibiofibular ligamentous or syndesmotic complex, the lateral collateral ligament (LCL) complex, and the deltoid ligament [19–23]. The LCL complex consists of three ligaments: ATF, CF, and PTF. The ATF ligament is located within the antero-

lateral joint capsule extending from the anteroinferior aspect of the lateral malleolus to the lateral talar neck (Fig. 2). The CF ligament is a cordlike structure that arises from the tip of the lateral malleolus and passes obliquely downward and posterior to insert at the posterolateral aspect of the calcaneus. It is extra-articular and closely associated with the inner sheath of the peroneal tendons. The PTF ligament is an intra-articular ligament that arises from the medial aspect of the distal fibula and passes almost horizontally to insert along the posterolateral tubercle of the talus [13–15,20,23,24].

The deltoid ligament or medial collateral ligament is composed of three superficial (tibionavicular, tibiospring, and tibiocalcaneal) and two deep (anterior and posterior tibiotalar) bands [17,20,22,25]. The deltoid ligament blends intimately with the tendon sheaths of the posterior tibial tendon and FHL and flexor digitorum longus tendons (Fig. 3). The anterior and posterior tibiofibular ligaments, the inferior transverse ligament, and the interosseous membrane form the distal tibiofibular syndesmosis ligamentous complex [17,20,22].

Approximately 85% of all ankle sprains are due to inversion forces and involve the LCL complex with a predictable sequence of injury involving first the ATF ligament, then the CF ligament, and finally the PTF ligament. Syndesmosis sprains are the second most prevalent (10%), and isolated medial sprains are third (5%) [17,20,26].

Most patients who had a lateral ankle sprain return to normal sport and daily living activity. Twenty

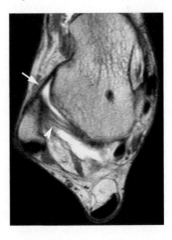

Fig. 2. Normal anatomy of the ATF and PTF ligaments of the LCL complex of the ankle. Axial T1-weighted MR arthrography shows the ATF ligament as a homogeneous low signal intensity structure (*arrow*) and the striated appearance of the PTF ligament (*arrowhead*) owing to the presence of fat interposed between its fascicles.

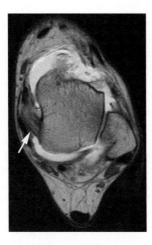

Fig. 3. Normal anatomy of the posterior tibiotalar fascicle of the deltoid ligament. Axial T1-weighted MR arthrography shows the normal striated posterior tibiotalar band of the deltoid ligament (*arrow*). Note the chronic tear of the ATF ligament.

percent to 40% of patients have residual pain, however, sufficient to limit or alter their activity [1, 26–28]. These cases constitute a diagnostic and therapeutic problem. Chronic pain presenting after lateral ankle sprains may be secondary to a variety of reasons, such as ankle instability, ankle soft tissue impingement syndromes, sinus tarsi syndrome, peroneal tendon lesions, or osteochondral lesions of the talar dome [14,22,27,29].

Ankle instability can be characterized as mechanical or functional. Frequent giving way without evidence of anatomic ligamentous incompetency commonly is referred to as *functional instability,* whereas the objective finding of ligament incompetency (mobility beyond the physiologic range of motion) is termed *mechanical instability.* The incidence of functional instability after ankle sprains has been reported to range from 15% to 60% and seems to be independent of the degree of severity of the initial injury. Mechanical instability is much less prevalent. Chronic ankle instability often is characterized by repeated episodes of giving way with asymptomatic periods between episodes [22,28–31].

Deltoid ligament injuries most commonly occur in association with lateral ligamentous pathology, a fibular fracture, or syndesmotic injuries [17,20,25]. Isolated ruptures of the deltoid ligament are rare, but can occur as a consequence of an eversion-lateral rotation injury. Contusions and partial tears of the deltoid ligament, particularly of its posterior tibiotalar component, frequently are associated with inversion sprains, in which the deep posterior fibers of the

medial deltoid ligament become crushed between the medial wall of the talus and the medial malleolus [20,32].

The incidence of syndesmosis sprains is probably higher than reported [19,25,28]. The mechanism of injury may be pronation and eversion of the foot combined with internal rotation of the tibia on a fixed foot. Syndesmosis injuries frequently are associated with eversion-type ankle fractures, particularly high fibular fractures (Maissoneuve) and rupture of the deltoid ligament. The presence of a syndesmosis sprain is a strong predictor for the likelihood of chronic ankle dysfunction [22,26,29].

Indications for the use of MR imaging may be limited to the evaluation of ligamentous injury in acute ankle injuries that show instability, stable acute injuries involving athletes or litigation, and repeated injuries or chronic ankle instability in patients in whom surgery is contemplated. MR imaging also has the advantage of depicting lesions often associated with ligamentous injuries, such as impingement syndromes, sinus tarsi syndrome, osteochondral lesions, and tendon tears [17,20,22].

Ankle ligaments are readily identified on MR images as low signal intensity structures joining adjacent bones usually delimited by contiguous high signal intensity fat. Heterogeneity and striation may be noted in some ligaments owing to the presence of fat interposed between their fascicles [15,17,22, 24,33,34].

The MR imaging criteria for the diagnosis of acute tears of the ankle ligaments include morphologic and signal intensity alterations within the ligament (primary signs) and around it (secondary signs). Primary signs of ligament tear include discontinuity, detachment, nonvisualization, or thickening of the ligament associated with increased intraligamentous signal intensity on T2-weighted images indicative of edema or hemorrhage. Secondary signs of acute ligament injury include extravasation of joint fluid into the adjacent soft tissues, joint effusion, and bone bruises. Fluid within the peroneal tendon sheath can be a secondary sign of acute CF ligament injury. In chronic tears, secondary signs disappear, and the ligament can show thickening, thinning, elongation, and irregular or wavy contour [13,15,17,20,21]. Avulsion injuries are diagnosed easily in the acute and chronic settings showing the bone fragment adjacent to an irregular lateral or medial malleolus.

It has been shown in a cadaver study that the normal ligaments of the LCL complex and of the syndesmotic complex are shown better on MR arthrography than on conventional MR imaging [1–3,16]. The improved visualization resulted from the contrast solution outlining more than one side of the ligament and because the ligaments were lifted away from the adjacent bone cortex.

MR arthrography is more sensitive and accurate than MR imaging in the evaluation of ligament tears [1–3,16]. The joint distention obtained with MR arthrography allows precise assessment of the thickness of the ligaments and their integrity at the insertion site, improving the diagnosis of acute and chronic tears (Fig. 4). Nonvisualization or extravasation of the contrast material anterior to the ATF ligament indicates tear of the ligament. A capacious anterior recess of the ankle joint allows the contrast agent to outline the anterior border of this capsular ligament, which represents capsular distention beyond the ligament and should not be confused with a

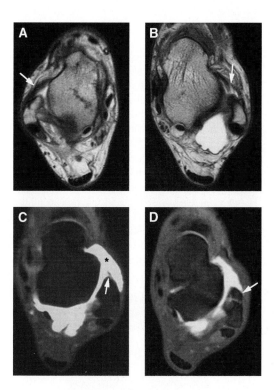

Fig. 4. Different appearances of chronic tears of the ATF ligament. (A) Axial T1-weighted MR arthrography shows irregular thickening and subtle waviness of the ATF ligament (arrow). (B) Axial T1-weighted MR arthrography shows focal disruption of the peroneal insertion of the ATF ligament (arrow). (C) Axial fat-suppressed T1-weighted MR arthrography shows disruption of the ATF ligament with small ligamentous end (arrow) and distention of the anterolateral recess (asterisk). (D) Axial T1-weighted MR arthrography reveals complete absence of the ATF ligament and avulsed osseous fragment (arrow) at the anteroinferior aspect of the lateral malleolus.

tear. Disruption of the CF ligament often results in extravasation of contrast material lateral to the ligament and in communication of the ankle joint with the peroneal tendon sheath, which is attached to the superficial surface of the ligament (Fig. 5). Accumulation of contrast material in the peroneal tendon sheath at MR arthrography is an indirect but specific sign of CF ligament injury. Extravasation of contrast material into the soft tissues posterior to the PTF ligament indicates a tear of this ligament [1–3,16].

Lee et al [16] found no advantage to using MR arthrography when visualization of the deltoid ligament was required. In the authors' experience, MR arthrography with optimal articular distention outlines the deep component of this ligament and improves evaluation of partial tears of the fascicles of this component (Fig. 6). MR arthrography allows a better assessment of tears of the syndesmotic ligaments, manifested as thickening, lack of visualization, or irregularity of these ligaments, and is helpful in the detection of associated lesions (Fig. 7) [16].

Treatment of lateral ankle ligamentous injuries is conservative and includes the *RICE* regimen (*r*est, *i*ce, *c*ompression, *e*levation) and early controlled motion with functional brace. Surgical management of acute ankle sprains is rarely indicated. Numerous surgical techniques have been described to correct ankle instability with an 80% to 90% success rate. Direct repair of the ATF and CF ligaments has a success rate similar to that for augmented reconstructions (tenodesis) [1,22,28,29].

Treatment of the deltoid ligament is controversial and depends of the associated lesions. Grade I and II lesions are managed conservatively. Isolated acute deltoid tear (grade III injuries), avulsion of the medial malleolus, and chronic deltoid sprains with lengthening ligament are treated surgically [22].

Partial isolated syndesmosis tears should be treated conservatively. A complete tear is managed by suture of the ligament and temporary fixation of the tibia and fibula with a syndesmosis screw, cerclage, or Kirschner wires [22,29].

Ankle impingement syndromes

Ankle impingement syndromes are painful entities caused by the friction of joint tissues, which is the cause and the effect of altered joint biomechanics. The leading causes of impingement lesions are posttraumatic ankle injuries, usually ankle sprains [35–38].

From the anatomic and clinical viewpoints, these syndromes are classified as anterolateral, anterior, anteromedial, posteromedial, and posterior [35–38]. Patient history and an adequate physical examination

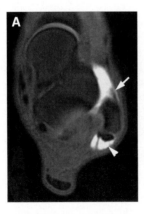
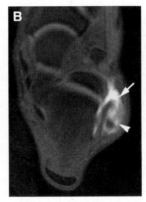
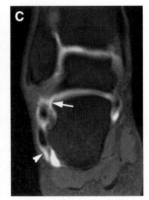

Fig. 5. Chronic tear of the ATF ligament and CF ligament. (*A* and *B*) Axial fat-suppressed T1-weighted MR arthrography shows disruption of the ATF (*arrow* in *A*) and CF ligaments (*arrow* in *B*). Note extravasation of fluid into the peroneal tendon sheath (indirect sign of the calcaneofibular ligament tear) (*arrowheads*). (*C*) Coronal fat-suppressed T1-weighted MR arthrography reveals discontinuity of the CF ligament (*arrow*) and communication of contrast with the peroneal tendon sheath (*arrowhead*).

can suggest a specific diagnosis in most cases. MR arthrography is the most useful imaging technique for detecting the soft tissue and osseous abnormalities present in these syndromes because it evaluates accurately the capsular recesses of the ankle [35–37].

Anterolateral impingement syndrome

Anterolateral impingement (ALI) is a relatively uncommon cause of chronic ankle pain produced by entrapment of abnormal soft tissue in the anterolateral gutter of the ankle [39–43]. It is estimated that approximately 3% of ankle sprains may lead to ALI. ALI is thought to occur subsequent to relatively minor trauma involving forced ankle plantar flexion and supination. Such trauma may result in tearing of

the anterolateral soft tissues and ligaments without substantial associated mechanical instability. Repeated microtrauma and soft tissue hemorrhage can result in synovial scarring, inflammation, and hypertrophy in the anterolateral gutter of the ankle, with subsequent soft tissue impingement. Wolin coined the term *meniscoid* owing to its resemblance at surgery to meniscal tissue [35–43].

Other contributing factors include hypertrophy of the inferior portion of the anteroinferior tibiofibular ligament and osseous spurs. First described by Bassett et al [44], a separate distal fascicle of the anteroinferior tibiofibular ligament is a common variant. It becomes pathologic when a tear of the anterior talofibular ligament results in anterolateral joint laxity. With increasing joint laxity, the talus

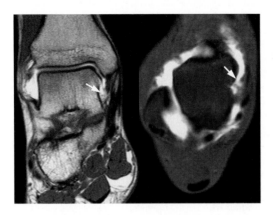

Fig. 6. Chronic tear of the deep component of the deltoid ligament. Coronal T1-weighted (*left*) and axial fat-suppressed T1-weighted (*right*) MR arthrography shows partial tear of the deltoid ligament involving the posterior tibiotalar fibers (*arrows*).

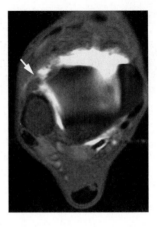

Fig. 7. Chronic syndesmosis sprain. Axial fat-suppressed T1-weighted image shows thickening and complete disruption of the inferior tibiofibular ligament (*arrow*).

extrudes anteriorly in dorsiflexion and comes into contact with the fascicle. Constant rubbing of the fascicle against the talus thickens the fascicle, which develops into an impinging lesion in the anterolateral gutter [44].

The clinical diagnosis of ALI can be established based on the combined presence of the following signs and symptoms: chronic ankle pain after an ankle sprain, anterolateral ankle joint tenderness, recurrent joint swelling, anterolateral pain with forced ankle dorsiflexion and eversion, pain during the single-leg squat, and lack of lateral ankle stability [37–39,42]. The clinical diagnosis of ALI is one of exclusion, however. Lesions producing similar symptoms have to be excluded before invasive treatment because similar symptoms can be attributed to peroneal tendon tears or subluxations, sinus tarsi syndrome, stress fractures, loose bodies, osteochondral lesions, bony impingement, and degenerative joint disease [35,37,38,41].

The presence of an abnormal soft tissue mass, hypointense on T1-weighted images and low or intermediate signal intensity on T2-weighted images, or a fibrous band in the anterolateral ankle gutter are MR imaging findings that suggest the diagnosis of ALI. The frayed margins of the torn ATF ligament should not be confused with the meniscoid lesion. Nevertheless, controversies exist regarding the accuracy of MR imaging in the diagnosis of ALI [39–42]. The evaluation of the anterolateral recess with conventional MR imaging is accurate only in the presence of substantial joint effusion [42].

MR arthrography is an accurate technique to assess the presence of soft tissue scarring in the anterolateral recess of the ankle and to determine its extent in patients with ALI before arthroscopy (Fig. 8). Robinson et al [41] in an arthroscopically controlled retrospective study found that MR arthrography assessment of the anterolateral soft tissues had an accuracy of 97%, sensitivity of 96%, and specificity of 100%. Accuracy was 100% with clinical anterolateral impingement. The absence of a recess of arthrographic fluid between the anterolateral soft tissues and the anterior surface of the fibula is another MR arthrography finding that suggests the diagnosis of ALI [41]. Nevertheless the identification of abnormal soft tissue itself does not imply the presence of clinical ALI. MR arthrography confirmation of anterolateral soft tissue abnormalities must be considered with the clinical findings [35,37].

Most patients with ALI respond to conservative therapy, including nonsteroidal anti-inflammatory drugs, rehabilitative physiotherapy, or local injection of steroids. If nonoperative treatment fails after 6 months, significant relief has been shown to be provided by arthroscopic débridement of hypertrophic synovial tissue in the anterolateral gutter [35–42].

Anterior impingement syndrome

Anterior impingement is seen more frequently in athletes subjected to repeated stress in dorsiflexion of the ankle, such as soccer players [35–37]. It is usually the result of impingement with trapping of soft tissues between a beaklike prominence at the anterior rim of the tibial plafond and a corresponding area over the apposing margin of the talus [35–37].

The cause and origin of anterior impingement are uncertain, and many factors are probably involved. It has been suggested that forced dorsiflexion results in repeated microtraumas on the tibia and talus leading to microfractures of trabecular bone or periosteal hemorrhage, which then heal with the formation of new bone. Another suggested mechanism in the etiology of these lesions is forced plantar flexion trauma, which causes capsular avulsion injury [35–37].

A classification for anterior ankle osteophytes has been developed, categorizing ankle spurs on the basis of spur size and the presence of associated arthritis. Lateral stress radiographs taken in maximum dorsiflexion may show physical impingement of the osteophytes [35,36]. MR arthrography is useful to assess the degree of cartilage damage, to delineate loose bodies, and to detect bone marrow edema and synovitis in the anterior capsular recess (Fig. 9) [35–37].

Conservative treatment, consisting of heel lifts, rest, modification of activities, and physical therapy, may be tried first. If there is persistent pain despite conservative treatment, arthroscopic or open resection of the spurs may be considered [35,36].

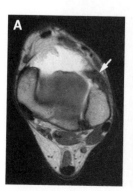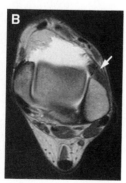

Fig. 8. Anterolateral impingement syndrome. (*A* and *B*) Serial axial T1-weighted MR arthrography reveals irregular soft tissue thickening in the anterolateral gutter (*arrows*).

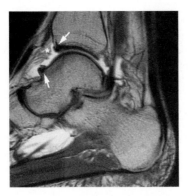

Fig. 9. Anterior impingement syndrome. Sagittal T1-weighted MR arthrography shows beaklike prominences at the anterior rim of the tibial plafond and over the opposed margin of the talus talar neck (*kissing lesion*) (*arrows*). Note synovitis in the anterior capsular recess of the tibiotalar joint (*asterisk*).

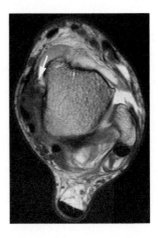

Fig. 10. Anteromedial impingement syndrome. Axial T1-weighted MR arthrography shows irregular soft tissue thickening in the anteromedial capsular recess (*arrow*) and thickening of the deep component of the deltoid ligament.

Anteromedial impingement syndrome

Anteromedial impingement is rarely an isolated condition, but is associated most commonly with an inversion mechanism of injury with lateral and medial ligamentous injury. It can be caused by a meniscoid lesion, represented by a mass of hyalinized connective tissue arising from a partially torn deep deltoid ligament or by a thickened anterior tibiotalar ligament. There also may be associated tibiotalar osteophytes. This thickened ligament or a meniscoid lesion, along with hypertrophic synovium, impinges on the anteromedial corner of the talus during dorsiflexion of the ankle [35–38,45].

Conventional MR imaging is not effective in detecting anteromedial impingement but can show a partially torn deep deltoid ligament. MR arthrography is the imaging method of choice, clearly defining the medial meniscoid lesion (Fig. 10), the thickened anterior tibiotalar ligament, and chondral or osteochondral associated lesions [35,45]. If conservative treatment fails, débridement of the impinging lesion by arthroscopic methods yields good clinical results [35,45].

Posteromedial impingement syndrome

Posteromedial impingement is a rare cause of ankle pain. It can occur after a severe ankle-inversion injury with the deep posterior fibers of the deltoid ligament becoming crushed between the talus and the medial malleolus. Initially the symptoms that predominate are from the lateral ligament tear. Inadequate healing of the contused deep posterior deltoid ligament fibers may lead, however, to chronic inflammation and hypertrophic fibrosis and metaplasia. In these cases, the fibrotic scar tissue may impinge between the medial wall of the talus and the posterior margin of the medial malleolus [35,46].

MR arthrography may show the thickened soft tissues (Fig. 11) and subchondral contusions of this entity [35]. This lesion generally cannot be fully appreciated arthroscopically via anterior portals in a stable ankle and requires a high index of suspicion and careful examination for the diagnosis to be made clinically. Arthroscopic surgery, via a posterior portal, or limited open surgery excision of the lesion is successful in resolving the pain [35,46].

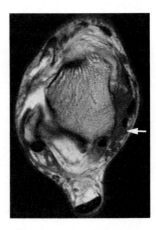

Fig. 11. Posteromedial impingement syndrome. Axial T1-weighted MR arthrography shows hypertrophic fibrotic tissue in the posteromedial aspect of the ankle (*arrow*).

Posterior impingement syndrome

Posterior ankle impingement (PAI) syndrome refers to a group of pathologic entities that result from repetitive or acute forced plantar flexion of the foot. [35,37,47] Ostrigonum syndrome, talar compression syndrome, and posterior block of the ankle are other names for the same syndrome. The mechanism of injury is the compression of the talus and the surrounding soft tissues between the tibia and the calcaneus, as a nut in a nutcracker, during plantar flexion of the foot. This syndrome is seen most commonly in classical ballet dancers, but it also can be seen in other sports [35,37,47,48].

The key factor in this syndrome is the anatomy of the posterior aspect of the ankle. The more common causes are the os trigonum (an accessory ossicle of the lateral tubercle that may persist unfused into adulthood in 7% of individuals), an elongated lateral tubercle termed a *Stieda process*, a downward sloping posterior lip of the tibia, the prominent posterior process of the calcaneus, and loose bodies. Synovitis of the FHL tendon sheath, the posterior synovial recess of the subtalar and tibiotalar joints, and the posterior intermalleolar ligament are possible soft tissue causes of impingement [35].

The presentation of the PAI syndrome can be as soft tissue inflammation, osseous injuries, or a combination of both. Osseous injuries include fracture, fragmentation, and pseudarthrosis of the os trigonum or lateral talar tubercle. The soft tissue changes associated are posterior ankle and subtalar synovitis and FHL tenosynovitis [35,37,47,48].

The diagnosis of PAI syndrome is based primarily on the patient's clinical history and physical examination results and is supported by findings on radio-graphs, CT, and MR imaging [37]. MR arthrography is useful in the assessment of PAI syndrome (Fig. 12). Findings include abnormal signal intensity in the lateral talar tubercle or os trigonum, consistent with bone marrow edema, which is believed to be the result of bone impaction and represents bone contusions or occult fractures. Inflammatory changes and synovitis in the posterior ankle soft tissues also can be found. The combined presence of marrow edema and posterior ankle synovitis may suggest the diagnosis of PAI [35,37].

The intermalleolar ligament often is not visualized on conventional MR imaging. MR arthrography improves the visualization of the intermalleolar ligament, which can readily be separated from the surrounding PTF ligament and the transverse inferior tibiofibular ligament [35,48,49].

The treatment of PAI syndrome is initially conservative. If conservative treatment fails, surgical excision of the osseous fragments, with potential release of the FHL tendon, may be indicated [21,35,36].

Cartilage lesions

Chondral lesions are common in the ankle joint. Symptoms of chondral lesions are mostly nonspecific, and clinical diagnosis is usually difficult. MR imaging is the best imaging method for the assessment of the articular cartilage [3,6]. Different MR imaging pulse sequences, including spin echo, gradient echo, fat-suppressed, and magnetization transfer contrast, have been used to study the articular cartilage. The reported sensitivities and specificities of different sequences vary. The following MR imaging grading is most commonly used: In grade I lesions,

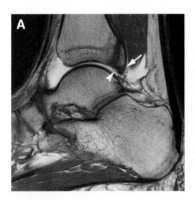
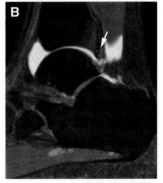

Fig. 12. Posterior impingement syndrome. (*A*) Sagittal T1-weighted MR arthrography shows a downward sloping deformity of the posterior lip of the tibia (*arrow*) after posterior malleolus fracture and focal chondral fraying (*arrowhead*). (*B*) Sagittal fat-suppressed, proton density–weighted spin-echo MR arthrography shows abnormal high signal intensity in the posterior malleolus (*arrow*).

MR images may show abnormal intrachondral signal with a smooth chondral surface and without alterations of the chondral thickness. Grade II lesions show mild surface irregularity with or without focal loss of less than 50% of the cartilage thickness. Severe surface irregularities with thinning of the cartilage thickness of more than 50% are present in grade III lesions. Grade IV lesions consist of complete loss of articular cartilage with denuded subchondral bone [3].

MR arthrography allows excellent delineation of the cartilage surface and provides good discrimination of higher grade cartilaginous lesions. MR arthrography can detect chondral lesions measuring 2 mm. Grade I chondral lesions have no surface abnormality and so may not be detected with MR arthrography [3,6,50].

Osteochondral lesions of the talus

Osteochondral lesion of the talus (OLT) is the accepted term for a variety of disorders including osteochondritis dissecans, osteochondral fracture, transchondral fracture, and talar dome fracture [51–53]. The primary lesional mechanism is a talar dome impaction owing to inversion injuries. These lesions typically involve medial or lateral aspects of the talar dome (Fig. 13). Medial lesions, which affect the posterior third of the talar border, are due to inversion injuries with plantar flexion of the foot and external rotation of the tibia (ie, impact between the posteromedial tibia and medial talar margin). Lateral lesions, which affect the anterior third of the talar border, are due to forced inversion and dorsiflexion

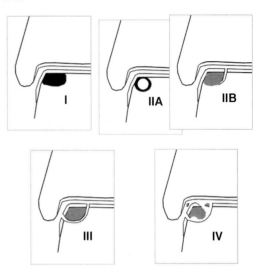

Fig. 14. Diagrams show classification of the osteochondral lesions of the talar dome.

of the foot (ie, impingement between the fibular styloid and the lateral margin of the talar dome). Medial and lateral aspects of the talar dome are involved in approximately 55% and 45% of the cases. Most lateral OLT appear thin and shallow, with the surface fragment greater in width than in depth. In contrast, medial OLT often have a deeper, crater-like appearance [36]. Clinical symptoms include exercise-related pain and less frequently sensations of clicking and catching and persistent swelling.

The classification introduced by Berndt and Harty is the most widely accepted staging system of osteochondral talar lesions [51–53]. This classification describes four stages depending on the integrity of the articular cartilage and the condition of the subchondral bone (Fig. 14). Stage I represents subchondral compression fracture, but the overlying articular cartilage remains intact. Stage II consists of a partially detached osteochondral fragment. In stage III, the osteochondral fragment is completely detached from the talus but is not displaced. In stage IV, the osteochondral fragment is detached and displaced, located away from the fracture site.

For a treatment decision, it is important to distinguish between stable and unstable lesions [19,34,53, 54]. In stable OLT, including stage I and most stage II lesions, conservative treatment is recommended. Surgical treatment is advocated for unstable lesions, including stage IV and most stage III OLT. A subset of stage II lesions, especially laterally located lesions, may be treated surgically. Conversely, a subset of stage III lesions, in particular lesions located in the

Fig. 13. Diagram shows the main locations of osteochondral lesions of the talus.

medial talar border, may be managed conservatively [19,34,53,54].

MR imaging is effective in characterizing all stages of OLT, but is most useful in the identification of radiographically occult OLT and the stratification of in situ lesions into stable and unstable subsets. MR imaging diagnosis of OLT instability has relied on the interface between the osteochondral fragment and the parent bone on T2-weighted images. A stable or healed osteochondral fragment is characterized by the lack of high signal intensity at the interface between the lesion and the parent bone. The presence of a high signal line on T2-weighted images at the talar interface with the osteochondral fragment is the most reliable sign of instability. This high signal intensity may represent granulation tissue or fluid. An interface that is hyperintense, but not as much as fluid, indicates the presence of fibrovascular granulation tissue or developing fibrocartilage. At this stage, the lesion is unstable, but has the capacity to heal after a period of non–weight bearing or internal fixation [53]. If the interface is isointense with fluid or associated with cystic-appearing areas at the base of a nondisplaced lesion, surgery is indicated. There is controversy as to the accuracy of MR imaging in assessing the stability of the osteochondral fragment, with one author reporting a low accuracy of 50%, believed to be caused in part by the inability to distinguish fluid from granulation tissue on T2-weighted sequences. MR arthrography provides a better depiction of the talar chondral surface and is useful in differentiating a stage II from a stage III lesion by documenting intra-articular communication of fluid around the lesion (Fig. 15), which aids in planning therapeutic arthroscopy [2,3,5,6].

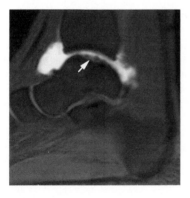

Fig. 15. Stage III osteochondral lesion. Sagittal T1-weighted MR arthrography shows a completely detached osteochondral fragment from the talus that remains located in the crater (*arrow*).

Surgical treatment consists of drilling of the lesion to improve perfusion or excision of the osteochondral fragment and débridement of the defect. Other surgical options include internal fixation of the fragment and bone grafting. In more advanced stage IV lesions, fragment excision and débridement of the defect is performed, whereas in earlier lesions (stage II and III), surgical treatment depends on the acute or chronic nature and size of the lesion [36,55,56].

Intra-articular loose bodies

Intra-articular loose bodies in the ankle joint may lead to impingement symptoms. Such bodies may consist of bone, cartilage, or both. Imaging is usually necessary to confirm the clinical diagnosis and to localize the intra-articular loose bodies before surgery. Radiographs are useful only when radiopaque intra-articular bodies are present. MR arthrography has been shown to be the best imaging technique for detecting osseous and cartilaginous loose bodies with a sensitivity of 86% and is significantly more sensitive than conventional MR imaging (Fig. 16) [57]. Air bubbles can mimic loose bodies at MR arthrography, but the distinction usually can be made by their nondependent position and typical appearance [2,5,6,7].

Synovial disorders

Synovial osteochondromatosis is an uncommon synovial metaplastic disorder that results in the formation of multiple ossified or calcified cartilaginous nodules [34,58,59]. It may arise in the synovium of joints, bursae, or tendon sheaths. Synovial osteochondromatosis is usually monarticular, with the knee involved most commonly. Clinical symptoms include pain, swelling, and locking of the affected joint. Radiographic findings depend on the presence of calcification or ossification within the nodules. Bone erosions are often seen. In one third of the cases, radiographs may be normal, owing to lack of ossification or calcification of the nodules. MR findings depend on the histologic composition of the nodules. Purely cartilaginous nodules are isointense with articular cartilage on all pulse sequences. Calcified nodules appear as signal void foci on all pulse sequences. By contrast, ossified nodules have a peripheral rim of low signal intensity on all pulse sequences and a central area of high T1 signal intensity corresponding to medullary fat [34,58,59]. MR arthrography in the absence of a significant joint effusion helps to confirm the diagnosis and to localize with precision the intra-articular bodies before surgery,

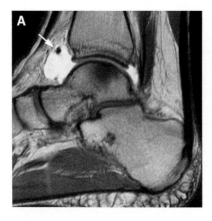
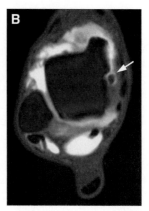

Fig. 16. Intra-articular loose bodies. (*A*) Sagittal T1-weighted MR arthrography shows osteochondral lesion of the talar dome and a small loose body in the anterior tibiotalar capsular recess (*arrow*). (*B*) Axial fat-suppressed T1-weighted MR arthrography shows a loose body in the anteromedial aspect of the ankle, anterior to the deep component of the deltoid ligament (*arrow*).

which is important because intra-articular bodies may be missed during arthroscopy (Fig. 17).

Pigmented villonodular synovitis (PVNS) is characterized by inflammatory proliferation of the synovium associated with deposits of hemosiderin [58]. It can be present in any joint, tendon sheath, or bursa but is seen most frequently in the knee, hip, ankle, and elbow. PVNS most often occurs in young to middle-aged adults and is more common in men. It may manifest as a focal mass or as a generalized lesion involving the entire joint space. Pressure erosions may be present in the diffuse form. These lesions manifest clinically as joint pain and swelling of long duration, and most are slowly progressive. PVNS has characteristic MR imaging features owing to the paramagnetic effect of hemosiderin,

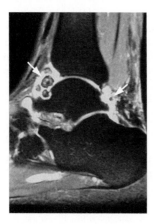

Fig. 17. Synovial osteochondromatosis. Sagittal fat-suppressed, proton density–weighted spin-echo MR arthrography shows multiple intra-articular osteochondral loose bodies (*arrows*).

which produces areas of low signal intensity on T1-weighted and T2-weighted images [58]. Although hemosiderin typically is present, however, it is not invariably detected by MR imaging. Occasionally, hyperintense areas may be seen within the lesion on T1-weighted images owing to the presence of fat or synovial hemorrhage. Joint effusion is often present, producing hypointense areas on T1-weighted images and hyperintense areas on T2-weighted images [58].

MR arthrography shows accurately the location and extension of PVNS lesions before treatment. Arthroscopic synovectomy may be indicated for a focal mass or for an inactive form of diffuse disease. Combined partial arthroscopic synovectomy and low-dose radiation therapy and arthroscopic or open synovectomy are indicated in the treatment of diffuse PVNS.

Arthrofibrosis is an abnormal proliferation of fibrous tissue in and around a joint that can occur after ankle trauma and particularly after immobilization for an ankle sprain or fracture, causing persistent pain, limitation of ankle motion, and disability [60]. The adhesions that form often lead to stiffness and abnormal joint contact pressures and predispose the joint to cartilage degeneration [60]. MR arthrography is useful in the diagnosis of this entity, showing marked diminution in the volume of the capsular space, retracted and irregular margins of the capsular insertions, and adhesions or fibrous bands.

Indirect MR arthrography

Intravenous administration of a standard dose of gadolinium followed by 5 to 10 minutes of light exercise can provide arthrogram-like images of the ankle joint [61–64]. This technique has been proposed as an alternative to direct MR arthrography. The main drawback of indirect MR arthrography is the lack of joint distention compared with direct MR arthrography. Another limitation of indirect MR arthrography is that juxta-articular structures, such as vessels, and the synovial membranes of bursae and tendon sheaths also show enhancement, which can lead to confusion with extravasation of contrast medium or the presence of abnormal joint recesses. Indirect MR arthrography may be useful in the evaluation of subtle cartilaginous defects, which are detected owing to enhancement of the cartilage defect and of the subchondral bone related to trabecular disruption and hyperemia [61–64].

In the assessment of OLT with indirect MR arthrography, contrast material enters the fragment-bone interface, indicating partial or complete detachment of the osteochondral fragment. Contrast can enter either from opacified synovial fluid or from the adjacent granulation tissue at the fragment-bone interface being hyperperfused. In either situation, the high T1 contrast at the interface is a sign of loosening (Fig. 18). Correlation of enhancement with T2-weighted signal improves accuracy [53,61–64].

Partial ligament tears may be identified by focal enhancement indicating hyperemia. Complete tears may be seen as enhanced joint fluid extending into the ligament defect (Fig. 19). Indirect MR arthrography also may be useful in the evaluation of ALI, outlining the inner aspect of the scarred ATF ligament [61–64].

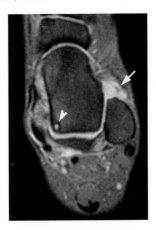

Fig. 19. Indirect MR ankle arthrography. Axial fat-suppressed T1-weighted MR arthrography shows contrast-enhanced fluid in the ankle joint that delineates a chronic tear of the ATF ligament (*arrow*) and small osteochondral lesion in the posteromedial aspect of the talar dome (*arrowhead*).

Indirect MR arthrography provides further assessment of extra-articular soft tissues of the ankle. Enhancement of extra-articular structures can highlight focal pathology, whereas lack of abnormal enhancement invariably indicates absence of disease in the region of interest. Enhancement about the plantar fascia is observed in patients with plantar fasciitis. With indirect MR arthrography, there is enhancement of fluid within the tendon sheath in the presence of tenosynovitis. Synovitis in the region of the tarsal tunnel is identified with enhancement around the posterior tibial nerve. Focal enhancement in the region of the sinus tarsi with indirect MR arthrography increases specificity of sinus tarsi pathology [61–64]. The overall advantage of indirect MR arthrography lies in gathering combined intra-articular and physiologic information.

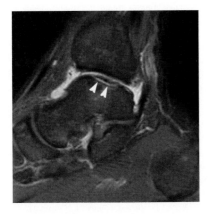

Fig. 18. Indirect MR ankle arthrography. Sagittal fat-suppressed T1-weighted MR arthrography shows contrast-enhanced fluid in the ankle joint around the osteochondral lesion of the talar dome (*arrowheads*), which indicates complete loosening of the osteochondral fragment.

Summary

MR arthrography has become an important tool for the assessment of a variety of ankle disorders. MR arthrography may facilitate the evaluation of patients with suspected intra-articular pathology in whom conventional MR imaging is not sufficient for an adequate diagnosis and be useful for therapy planning. MR arthrography is valuable in the evaluation of ligamentous injuries, impingement syndromes, cartilage lesions, OLT, loose bodies, and several synovial joint disorders. Indirect MR arthrography is a useful adjunct to conventional MR imaging and may be preferable to direct MR arthrography in cases in which an invasive procedure is contraindicated or when fluoroscopy is not available.

References

[1] Chandnani VP, Harper MT, Ficke JR, et al. Chronic ankle instability: evaluation with MR arthrography, MR imaging, and stress radiography. Radiology 1994; 192:189–94.

[2] Helgason JW, Chandnani VP. MR arthrography of the ankle. Radiol Clin North Am 1998;36:729–38.

[3] Kramer J, Recht MP. MR arthrography of the lower extremity. Radiol Clin North Am 2002;40:1121–32.

[4] Elentuck D, Palmer WE. Direct magnetic resonance arthrography. Eur Radiol 2004;14:1956–67.

[5] Grainger AJ, Elliott JM, Campbell RS, Tirman PF, Steinbach LS, Genant HK. Direct MR arthrography: a review of current use. Clin Radiol 2000;55:163–76.

[6] Peh WC, Cassar-Pullicino VN. Magnetic resonance arthrography: current status. Clin Radiol 1999;54: 575–87.

[7] Steinbach LS, Palmer WE, Schweitzer ME. Special focus session: MR arthrography. Radiographics 2002; 22:1223–46.

[8] Brown RR, Clarke DW, Daffner RH. Is a mixture of gadolinium and iodinated contrast material safe during MR arthrography? AJR Am J Roentgenol 2000;175: 1087–90.

[9] Haller J, Resnick D, Sartoris D, Mitchell M, Howard B, Gilula L. Arthrography, tenography, and bursography of the ankle and foot. Clin Podiatr Med Surg 1988;5:893–908.

[10] Pfirrmann CW, Zanetti M, Hodler J. Joint magnetic resonance imaging: normal variants and pitfalls related to sports injury. Radiol Clin North Am 2002;40: 167–80.

[11] Robbins MI, Anzilotti Jr KF, Katz LD, Lange RC. Patient perception of magnetic resonance arthrography. Skeletal Radiol 2000;29:265–9.

[12] Schulte-Altedorneburg G, Gebhard M, Wohlgemuth WA, et al. MR arthrography: pharmacology, efficacy and safety in clinical trials. Skeletal Radiol 2003;32: 1–12.

[13] Schneck CD, Mesgarzadeh M, Bonakdarpour A, Ross GJ. MR imaging of the most commonly injured ankle ligaments: part I. normal anatomy. Radiology 1992; 184:499–506.

[14] Cardone BW, Erickson SJ, Den Hartog BD, Carrera GF. MRI of injury to the lateral collateral ligamentous complex of the ankle. J Comput Assist Tomogr 1993; 17:102–7.

[15] Erickson SJ, Smith JW, Ruiz ME, et al. MR imaging of the lateral collateral ligament of the ankle. AJR Am J Roentgenol 1991;156:131–6.

[16] Lee SH, Jacobson J, Trudell D, Resnick D. Ligaments of the ankle: normal anatomy with MR arthrography. J Comput Assist Tomogr 1998;22:807–13.

[17] Bencardino JT, Rosenberg ZS. Normal variants and pitfalls in MR imaging of the ankle and foot. Magn Reson Imaging Clin N Am 2001;9:447–63.

[18] Rosenberg ZS, Bencardino J, Mellado JM. Normal variants and pitfalls in magnetic resonance imaging of the ankle and foot. Top Magn Reson Imaging 1998; 9:262–72.

[19] Bencardino J, Rosenberg ZS, Delfaut E. MR imaging in sports injuries of the foot and ankle. Magn Reson Imaging Clin N Am 1999;7:131–49.

[20] Cheung Y, Rosenberg ZS. MR imaging of ligamentous abnormalities of the ankle and foot. Magn Reson Imaging Clin N Am 2001;9:507–31.

[21] Dunfee WR, Dalinka MK, Kneeland JB. Imaging of athletic injuries to the ankle and foot. Radiol Clin North Am 2002;40:289–312.

[22] Narvaez JA, Cerezal L, Narvaez J. MRI of sports-related injuries of the foot and ankle: part 1. Curr Probl Diagn Radiol 2003;32:139–55.

[23] Beltran J, Munchow AM, Khabiri H, Magee DG, McGhee RB, Grossman SB. Ligaments of the lateral aspect of the ankle and sinus tarsi: an MR imaging study. Radiology 1990;177:455–8.

[24] Muhle C, Frank LR, Rand T, et al. Collateral ligaments of the ankle: high-resolution MR imaging with a local gradient coil and anatomic correlation in cadavers. Radiographics 1999;19:673–83.

[25] Klein MA. MR imaging of the ankle: normal and abnormal findings in the medial collateral ligament. AJR Am J Roentgenol 1994;162:377–83.

[26] Gerber JP, Williams GN, Scoville CR, Arciero RA, Taylor DC. Persistent disability associated with ankle sprains: a prospective examination of an athletic population. Foot Ankle Int 1998;19:653–60.

[27] DIGiovanni BF, Fraga CJ, Cohen BE, Shereff MJ. Associated injuries found in chronic lateral ankle instability. Foot Ankle Int 2000;21:809–15.

[28] Peters JW, Trevino SG, Renstrom PA. Chronic lateral ankle instability. Foot Ankle 1991;12:182–91.

[29] Renstrom PA. Persistently painful sprained ankle. J Am Acad Orthop Surg 1994;2:270–80.

[30] Colville MR. Surgical treatment of the unstable ankle. J Am Acad Orthop Surg 1998;6:368–77.

[31] Karlsson J, Lansinger O. Lateral instability of the ankle joint. Clin Orthop 1992;276:253–61.

[32] van Dijk CN, Bossuyt PM, Marti RK. Medial ankle pain after lateral ligament rupture. J Bone Joint Surg Br 1996;78:562–7.

[33] Schneck CD, Mesgarzadeh M, Bonakdarpour A. MR imaging of the most commonly injured ankle ligaments: part II. ligament injuries. Radiology 1992;184: 507–12.

[34] Rosenberg ZS, Beltran J, Bencardino JT, From the RSNA Refresher Courses, Radiological Society of North America. MR imaging of the ankle and foot. Radiographics 2000;20(Spec No):S153–79.

[35] Cerezal L, Abascal F, Canga A, et al. MR imaging of ankle impingement syndromes. AJR Am J Roentgenol 2003;181:551–9.

[36] Narvaez JA, Cerezal L, Narvaez J. MRI of sports-related injuries of the foot and ankle: part 2. Curr Probl Diagn Radiol 2003;32:177–93.

[37] Robinson P, White LM. Soft-tissue and osseous impingement syndromes of the ankle: role of imaging in diagnosis and management. Radiographics 2002;22:1457–69.

[38] Umans H. Ankle impingement syndromes. Semin Musculoskelet Radiol 2002;6:133–9.

[39] Farooki S, Yao L, Seeger LL. Anterolateral impingement of the ankle: effectiveness of MR imaging. Radiology 1998;207:357–60.

[40] Jordan III LK, Helms CA, Cooperman AE, Speer KP. Magnetic resonance imaging findings in anterolateral impingement of the ankle. Skeletal Radiol 2000;29:34–9.

[41] Robinson P, White LM, Salonen DC, Daniels TR, Ogilvie-Harris D. Anterolateral ankle impingement: MR arthrographic assessment of the anterolateral recess. Radiology 2001;221:186–90.

[42] Rubin DA, Tishkoff NW, Britton CA, Conti SF, Towers JD. Anterolateral soft-tissue impingement in the ankle: diagnosis using MR imaging. AJR Am J Roentgenol 1997;169:829–35.

[43] Wolin I, Glassman F, Sideman S, Levinthal DH. Internal derangement of the talofibular component of the ankle. Surg Gynecol Obstet 1950;91:193–200.

[44] Bassett III FH, Gates III HS, Billys JB, Morris HB, Nikolaou PK. Talar impingement by the anteroinferior tibiofibular ligament: a cause of chronic pain in the ankle after inversion sprain. J Bone Joint Surg Am 1990;72:55–9.

[45] Robinson P, White LM, Salonen D, Ogilvie-Harris D. Anteromedial impingement of the ankle: using MR arthrography to assess the anteromedial recess. AJR Am J Roentgenol 2002;178:601–4.

[46] Paterson RS, Brown JN. The posteromedial impingement lesion of the ankle: a series of six cases. Am J Sports Med 2001;29:550–7.

[47] Bureau NJ, Cardinal E, Hobden R, Aubin B. Posterior ankle impingement syndrome: MR imaging findings in seven patients. Radiology 2000;215:497–503.

[48] Fiorella D, Helms CA, Nunley JA. The MR imaging features of the posterior intermalleolar ligament in patients with posterior impingement syndrome of the ankle. Skeletal Radiol 1999;28:573–6.

[49] Rosenberg ZS, Cheung YY, Beltran J, Sheskier S, Leong M, Jahss M. Posterior intermalleolar ligament of the ankle: normal anatomy and MR imaging features. AJR Am J Roentgenol 1995;165:387–90.

[50] Schmid MR, Pfirrmann CW, Hodler J, Vienne P, Zanetti M. Cartilage lesions in the ankle joint: comparison of MR arthrography and CT arthrography. Skeletal Radiol 2003;32:259–65.

[51] Berndt AL, Harty M. Transchondral fractures (osteochondritis dissecans) of the talus. Am J Orthop 1959;41A:988–1020.

[52] Anderson IF, Crichton KJ, Grattan-Smith T, Cooper RA, Brazier D. Osteochondral fractures of the dome of the talus. J Bone Joint Surg Am 1989;71:1143–52.

[53] Beltran J, Shankman S. MR imaging of bone lesions of the ankle and foot. Magn Reson Imaging Clin N Am 2001;9:553–66.

[54] De Smet AA, Fisher DR, Burnstein MI, Graf BK, Lange RH. Value of MR imaging in staging osteochondral lesions of the talus (osteochondritis dissecans): results in 14 patients. AJR Am J Roentgenol 1990;154:555–8.

[55] Assenmacher JA, Kelikian AS, Gottlob C, Kodros S. Arthroscopically assisted autologous osteochondral transplantation for osteochondral lesions of the talar dome: an MRI and clinical follow-up study. Foot Ankle Int 2001;22:544–51.

[56] Schimmer RC, Dick W, Hintermann B. The role of ankle arthroscopy in the treatment strategies of osteochondritis dissecans lesions of the talus. Foot Ankle Int 2001;22:895–900.

[57] Brossmann J, Preidler KW, Daenen B, et al. Imaging of osseous and cartilaginous intraarticular bodies in the knee: comparison of MR imaging and MR arthrography with CT and CT arthrography in cadavers. Radiology 1996;200:509–17.

[58] Weishaupt D, Schweitzer ME, Alam F, Karasick D, Wapner K. MR imaging of inflammatory joint diseases of the foot and ankle. Skeletal Radiol 1999;28:663–9.

[59] Boles CA, Ward Sr WG. Loose fragments and other debris: miscellaneous synovial and marrow disorders. Magn Reson Imaging Clin N Am 2000;8:371–90.

[60] Amendola A, Petrik J, Webster-Bogaert S. Ankle arthroscopy: outcome in 79 consecutive patients. Arthroscopy 1996;12:565–73.

[61] Bergin D, Schweitzer ME. Indirect magnetic resonance arthrography. Skeletal Radiol 2003;32:551–8.

[62] Schweitzer ME, Natale P, Winalski CS, Culp R. Indirect wrist MR arthrography: the effects of passive motion versus active exercise. Skeletal Radiol 2000;29:10–4.

[63] Vahlensieck M, Peterfy CG, Wischer T, et al. Indirect MR arthrography: optimization and clinical applications. Radiology 1996;200:249–54.

[64] Zoga AC, Schweitzer ME. Indirect magnetic resonance arthrography: applications in sports imaging. Top Magn Reson Imaging 2003;14:25–33.

RADIOLOGIC CLINICS
of North America

Radiol Clin N Am 43 (2005) 709 – 731

Wrist MR Arthrography: How, Why, When

Luis Cerezal, MD[a],*, Faustino Abascal, MD[a], Roberto García-Valtuille, MD[a], Francisco del Piñal, MD[b]

[a]Department of Radiology, Instituto Radiológico Cántabro, Clínica Mompía, Mompía, Cantabria 39109, Spain
[b]Department of Private Hand–Wrist and Plastic–Reconstructive Surgery and Hand Surgery, Mutua Montañesa, Calderón de la Barca 16-entlo, Santander 39002, Spain

MR imaging has been used in the evaluation of a wide spectrum of joint disorders. Its multiplanar capabilities and refined tissue contrast allow detailed assessment of osseous and soft tissue pathology. MR imaging of the wrist frequently represents a diagnostic challenge for radiologists because of the complex anatomy of the wrist joint, the small size of its components, and little known pathologic conditions. MR arthrography combines the advantages of conventional MR imaging and arthrography by improving the visualization of small intra-articular abnormalities. MR arthrography of the wrist is a mildly invasive imaging technique, however, and should not be performed indiscriminately. This article reviews the current role of MR arthrography in the evaluation of wrist joint disorders taking into account the relevant aspects of anatomy, techniques, and applications.

How to perform wrist MR arthrography

Triple- (midcarpal, radiocarpal, and distal radioulnar joint [DRUJ]), double- (radiocarpal and midcarpal or radiocarpal and DRUJ), and single- (radiocarpal) compartment MR arthrography have been used in the wrist [1–11]. Intra-articular injection of a contrast agent is generally performed under fluoroscopic guidance. Sonographic, CT, or MR imaging guidance may be also used [12–14].

Multiple sites can be selected to successfully distend the midcarpal and radiocarpal joints (Fig. 1) [15]. The injection site of choice should be on the side of the patient's wrist opposite the symptoms to help distinguish iatrogenic spill into the dorsal soft tissues from a true capsular disruption. Injection sites for the midcarpal compartment include the distalmost scaphocapitate and triquetrohamate spaces [15]. Injection should continue until the contrast is readily visualized in the capitolunate joint space. In normal arthrograms, contrast flows into both the scapholunate (SL) and lunotriquetral (LT) spaces. The intrinsic ligaments at the proximal margins of the scaphoid, lunate, and triquetral bones arrest the proximal flows of contrast, preventing communication with the radiocarpal compartment. For radiocarpal joint ulnar-sided injections, the needle should be directed to the proximal edge of the triquetrum at the pisiform radial margin. For radial-sided injections, the needle should be directed to the radioscaphoid space away from the SL joint. Because of the natural volar tilt of the distal radius, a slight angulation of the imaging intensifier in the cranial direction facilitates better profiling of the radioscaphoid space. This prevents the needle-tip from striking the dorsal lip of the radius, which frequently overlays the radioscaphoid space on a true posteroanterior projection. Alternatively, the needle-tip can be directed into the proximal scaphoid at the margin of the radial styloid. Contrast generally fills the dorsal recess of the radiocarpal joint; however, care must be taken not to inject the area from the scaphoid distal to the scaphoid tubercle or inadvertent filling of the midcarpal joint may occur [15].

* Corresponding author.
 E-mail address: lcerezal@mundivia.es (L. Cerezal).

doi:10.1016/j.rcl.2005.02.004

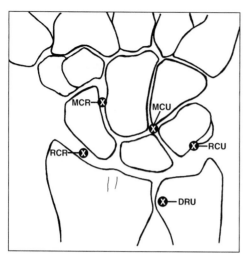

Fig. 1. Diagram illustrating the injection sites for wrist joint MR arthrography. DRU, distal radioulnar joint site; MCR, midcarpal radial site; MCU, midcarpal ulnar site; RCR, radiocarpal radial site; RCU, radiocarpal ulnar site.

The radiocarpal joint communicates with the pisotriquetral joint in 34% to 70% of patients.

The DRUJ surrounds the head of the ulna and extends to, but does not include, the ulnar aspect of the radius. Placement of the needle at the midpoint of the DRUJ space can result in an extra- as well as an intra-articular injection. The needle-tip should be directed toward the head of the ulna near its radial margin. After the needle touches the ulnar head, it should be slightly directed radially to advance deeper into the joint space thus stabilizing the needle [15]. It is important that the injected contrast show the fovea at the base of the ulnar styloid to establish where there is a defect in the ulnar attachment of the triangular fibrocartilage (TFC).

Triple-compartment wrist arthrography is performed using the following technique: first, the patient is positioned supine on the fluoroscopic table with the wrist in neutral rotation and slight volar flexion. Traction on the patient's hand is helpful, but not essential, for a successful injection. After skin preparation with a povidone–iodine solution, a 25-gauge needle is inserted under fluoroscopic guidance directly through the skin from a dorsal approach. Small-bore needles (25-gauge) used for wrist arthrography have the advantage of minimizing tissue trauma and postinjection leakage of joint fluid. The needle position is verified by a test injection of a small amount of iodinated contrast agent (approximately 1–2 mL) before administering diluted gadolinium. If the needle is intra-articular, the contrast solution flows away from the needle-tip drawing the

capsular recesses. Subsequently, a solution of 0.1-mL gadolinium diluted in 20 mL of a solution composed of 10 mL of saline, 5 mL of iodinated contrast material, and 5 mL of lidocaine 1% is injected [16]. Triple-compartment MR arthrography is performed first with the injection into the midcarpal joint [17]. A total volume of 3 mL to 4 mL of solution is injected. If communication with the radiocarpal joint is present, an additional 3 mL to 4 mL of solution is injected. If communication with the radiocarpal joint and the DRUJ occurs, a further additional 1 mL to 2 mL of solution is added, for a total of 7 mL to 9 mL. If no communication is present, the radiocarpal joint and the DRUJ are sequentially injected with 3 mL to 4 mL and 1 mL to 2 mL of the solution, respectively.

Saline solution may be injected as arthrographic contrast material; however, the imaging characteristics of intra-articular gadolinium provide specific advantages over saline [12–14]. Saline within the joint is similar to a high-signal joint effusion seen on T2-weighted sequences and, although fluid collections in the vicinity of the joint are evident with T2-weighted images, it is not possible to determine whether the fluid collections are separate from the joint space or if they have occurred as a result of the saline injection.

Radiocarpal joint injection can be performed easily without imaging guidance using recognized palpable landmarks and avoiding the use of iodinated contrast medium and ionizing radiation [18]. This is especially useful in cases of limited access to a fluoroscopic suite. The injection site represents the arthroscopic 3-4 portal, an anatomic sulcus located between the extensor pollicis longus and index finger extensor digitorum communis tendon, just distal to Lister's tubercle, which provides the initial reference. The extensor pollicis longus tendon is easily palpated just ulnar to Lister's tubercle. The index finger extensor digitorum communis tendon is palpated just ulnar to the extensor pollicis longus tendons. Palpation at the 3-4 portal reveals a depression that is more defined when the patient alternatively extends the thumb (contracting the extensor pollicis longus tendon) and fingers (contracting the extensor digitorum communis tendons). With deep palpation during radiocarpal flexion and extension, the level of the dorsal lip of the radius can be estimated by feeling the mobile carpus as it moves on the fixed radius. The skin entry site is approximately 0.5 cm distal to the dorsal lip of the radius so that needle entry is angulated parallel to the distal radial articular surface (approximately 10–15°). The injection needle typically requires insertion to a depth of 0.5 cm to 1.5 cm [18]. During injection, radiocarpal distention is usually visible and ballotable, particularly in the so-called "anatomic snuffbox" region. Although injections can be performed successfully without fluoroscopy guidance, it is advisable to have it ready available as a useful and reassuring tool when initially learning the injection procedure [18].

The current authors perform triple-compartment MR arthrography with fluoroscopic guidance in patients who have chronic pain of unclear origin or instability syndromes of the wrist. In cases of suspected TFC complex (TFCC) or intrinsic ligaments lesions, the authors prefer to practice conventional MR imaging complemented with radiocarpal MR arthrography performed directly in the MR suite, based on anatomic landmarks without fluoroscopic guidance. Only in selected cases, when suspected lesion of ulnar attachment of TFC exists at conventional MR imaging, do they perform double-compartment (radiocarpal and DRUJ) MR arthrography with fluoroscopic guidance. MR images should be obtained shortly after conventional arthrography to minimize absorption of contrast and guarantee the desired capsular distention [12–14].

MR imaging evaluation of the wrist has lagged behind that of other larger joints because of the technical limitations of spatial resolution and the signal-to-noise ratio when imaging the small structures of the wrist [19,20]. Many of the larger ligaments around the wrist are no greater than 1 mm to 2 mm thick. High-resolution MR imaging is essential in evaluating normal features and pathologic conditions of the wrist. Recent advances in MR imaging coil design have dramatically changed the capabilities of studying small hand and wrist structures. Adapted wrist coils, such as quadrature, phased array, and special microscopy coils designed for high-resolution wrist imaging, optimize spatial and contrast resolution for the small field of view (3–10 cm) and thin slice thickness (1–2 mm) [19].

Imaging in the coronal, axial, and sagittal planes should be performed in comprehensive wrist evaluation. Demonstration of the intrinsic intercarpal ligaments and TFCC are best identified with the coronal plane [21].

Numerous MR imaging sequences are currently available for wrist assessment [22]. T1-weighted spin-echo sequences with and without fat suppression maximize the signal intensity of contrast solution. Fat suppression is crucial in MR arthrography because fat and gadolinium have similar signal intensities in T1-weighted images, which makes diagnosis difficult. Fat suppression selectively decreases the signal from fat and the signal from the contrast solution is preserved, thereby confirming or excluding extra-articular contrast material. At least one fat-suppressed T2-weighted sequence should be performed for the detection of subtle bone marrow edema and extra-articular fluid collections. Three-dimensional gradient echo images are helpful in the assessment of TFCC and ligaments [23–28]. Overall, the choice of sequence depends on the preferences of the radiologist and sequence availability.

As with any invasive procedure, it is important to consider the potential risks of wrist MR arthrography [9]. The use of iodinated contrast material to confirm intra-articular needle placement carries a small risk of reaction. Slight joint pain is relatively common but usually disappears within the first few hours after puncture. Vasovagal reactions are rare and managed easily in the fluoroscopic suite so that routine administration of prophylactic atropine before arthrography is unnecessary. The most serious complication of arthrography is joint infection. Its incidence is rare. Currently, no side effects from intra-articular gadolinium solution use have been reported [9].

In terms of pitfalls, the most common are extra-capsular contrast injection and extravasation of contrast material outside the joint, possibly as a result of overdistention, which can be mistaken for capsular disruption. Inadvertent use of undiluted gadolinium leads to T1 and T2 shortening and very little fluid appearing in signal. Air bubbles introduced during injection may mimic loose bodies, although generally they are in nondependent positions [12–14].

Why and when to perform wrist MR arthrography

Triangular fibrocartilage complex

In 1981, Palmer and Werner [29] introduced the term "triangular fibrocartilage complex" to describe the complex of soft tissues interposed between the distal part of the ulna and the ulnar carpus. In most descriptions the TFCC is composed of the TFC proper, the meniscus homologue, the ulnar collateral ligament (UCL), the dorsal and volar radioulnar ligaments, the subsheath of the extensor carpi ulnaris tendon or infratendinous extensor retinaculum, and the ulnocarpal ligaments (Fig. 2) [29,30]. Proximally, the TFCC originates at the ulnar aspect of the sigmoid notch of the radius extending toward the ulna pole and into the fovea at the base of the ulnar styloid. Two types of ulna attachments are observed. The most common is composed of two striated fascicles: one inserted at the base of the styloid and the other at the styloid tip. A less common insertion is

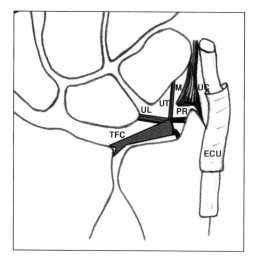

Fig. 2. Anatomy and schematic drawing of the TFCC. ECU, extensor carpi ulnaris tendon; M, meniscus homologue; PR, prestyloid recess; TFC, triangular fibrocartilage; UC, ulnar collateral ligament; UL, ulnolunate; UT, ulnotriquetral ligament.

a broad-based striated fascicle attachment along the entire length of the ulnar styloid [27]. Distally, the TFCC extends into the hamate, triquetrum, and base of the fifth metacarpal [29,30], and distally, it is joined by fibers of the UCL.

The TFC is a semicircular, fibrocartilaginous, bi-concave structure interposed between the ulnar dome and the ulnar aspect of the carpus. TFC thickness is inversely proportional to the ulnar variance: ulna-minus wrists have thick TFC and vice versa. The peripheral attachment of the TFC is ~5 mm thick, thinning at the center, and then narrowing to less than 2 mm [30].

From the anterior edge of the TFC, two groups of longitudinally-oriented collagen fibers emerge: the ulnotriquetral ligament, which runs distally and into the volar aspect of the triquetrum, and the ulnolunate ligament, which runs obliquely and then distally inserted into the lunate (Fig. 3) [29,30]. The thick, strong peripheral margins of the TFC, which are composed of lamellar collagen, are often referred to as the dorsal and volar radioulnar ligaments.

The meniscus homologue is an ill-defined region of well-vascularized and loose connective tissue on the volar side of the wrist. It has a common origin with the dorsal radioulnar ligament on the dorsoulnar corner of the radius. The meniscus homologue inserts directly into the triquetrum and partially or completely separates the pisotriquetral from the radiocarpal joint [30].

The extensor carpi ulnaris tendon is located within a dorsal notch in the distal ulna. The extensor carpi ulnaris subsheath makes a significant contribution to

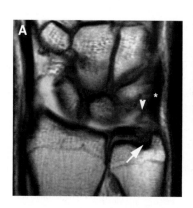
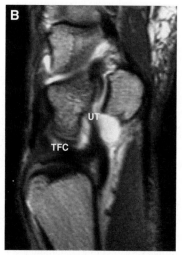

Fig. 3. Anatomy of the TFCC. (A) Coronal T1-weighted MR arthrogram showing a normal TFCC. The TFC appears as a low-signal intensity triangular structure. Note the striated appearance of ulnar insertion of the TFC (*arrow*). The meniscus homologue appears as an ill-defined low-signal region on the volar ulnar side of the wrist (*arrowhead*). The extensor carpi ulnaris tendon subsheath fuses with the dorsal aspect of the TFC (*asterisk*). (B) Sagittal T1-weighted MR arthrogram image showing the normal discoid appearance of the TFC and a longitudinally-oriented ulnotriquetral ligament, which courses distally until its insertion into the volar aspect of the triquetrum. TFC, triangular fibrocartilage; UT, ulnotriquetral ligament.

the stabilization of the dorsal TFCC because some of its fibers fuse to it [30].

The blood supply of the TFCC originates from the ulnar artery (through the radiocarpal branches) and the anterior interosseous artery (through the dorsal and volar branches). These vessels only peripheraly penetrate 10% to 40% of the TFCC, and the central and radial portions are avascular [31–33]. This pattern of supply has direct implications to the healing potential following injury of the TFC and the radioulnar ligaments, with peripheral ulnar-sided detachments demonstrating a superior healing capacity following repair when compared with radial-sided detachments.

The TFCC has three main functions. (1) it is the major stabilizer of the DRUJ and one of the stabilizers of the ulnar carpus. The primary function of radioulnar ligaments is to prevent volar and dorsal subluxation at the DRUJ. (2) The TFCC is also a load-bearing structure between the ulnar head and lunate and triquetrum. And (3) the ulnolunate and ulnotriquetral ligaments prevent volar subluxation of the ulnar carpus [29,30].

Triangular fibrocartilage complex tears

TFCC lesions may be variable in their extent of involvement. They may be confined to TFC or involve one or more components of the TFCC. Degenerative and traumatic tears of the TFC may occur. Degenerative tears are more common than traumatic tears. The incidence of central degenerative tears is age-related. According to Mikic [34], degeneration begins in the third decade and progressively increases in frequency and severity in subsequent

decades. The changes are more frequent and more intense on the ulnar surface, and they are always situated in the central part of the TFC [35,36].

In 1989, Palmer [37] proposed a classification system for TFCC tears that divided these injuries into two categories: traumatic (class I) and degenerative (class II). Traumatic tears are more common in younger patients and are subclassified according to the site of TFCC involvement (Fig. 4). Class IA, the central perforation, represents a tear or perforation of the horizontal portion of the TFCC, usually occurring as a 1- to 2-mm slit, and located 2 to 3 mm medial to the radial attachment of the TFCC. Class IB represents a traumatic avulsion of the TFCC from its insertion site into the distal portion of the ulna, sometimes with an associated fracture at the base of the ulnar styloid. Class IC represents distal avulsion of the TFCC at its site of attachment to the lunate or triquetrum reflective of a tear of the ulnolunate or ulnotriquetral ligaments. A class ID lesion represents an avulsion of the TFCC from its attachment to the radius at the distal aspect of the sigmoid notch, which may be associated with an avulsion fracture of this region.

The degenerative types reflect the progressive stages of ulnar impaction syndrome and are subclassified according to the degree of involvement of structures on the ulnar side [37,38], highlighting the progressive nature of these injuries (Fig. 5). Class IIA injuries represent TFC wear from the undersurface, occurring in the central horizontal portion, without perforation. Class IIB includes TFC wear with associated lunate or chondromalacia. Class IIC injuries represent TFC perforation with lunate or ulnar

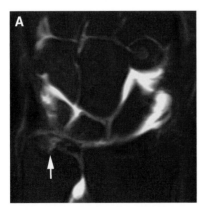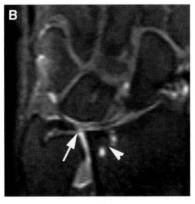

Fig. 4. Traumatic tears of the TFCC. (*A*) Coronal fat-suppressed T1-weighted radiocarpal MR arthrogram image showing traumatic avulsion of the ulnar attachment of the TFC (Palmer class IB lesion) with contrast leakage into the distal radioulnar joint (*arrow*). (*B*) Coronal fat-suppressed T1-weighted radiocarpal MR arthrogram image revealing an avulsion of the TFC from its attachment to the radius (*arrow*), associated with an avulsion fracture at the distal aspect of the sigmoid notch (*arrowhead*) (Palmer class ID lesion).

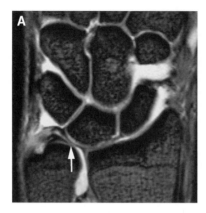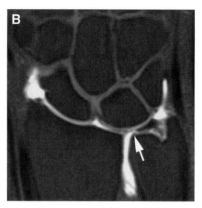

Fig. 5. Degenerative tears of the TFCC. (*A*) Palmer class IIB lesion (ulnar impaction syndrome). Coronal T2*-weighted MR arthrogram image revealing a degenerative partial tear of the proximal surface of the central portion of the TFC and thinning without perforation (*arrow*). Note the existence of positive ulnar variance. (*B*) Palmer class IIC lesion (ulnar impaction syndrome). Coronal fat-suppressed T1-weighted MR arthrogram image showing a central TFC perforation with contrast material communication between the radiocarpal and distal radioulnar compartments (*arrow*).

chondromalacia. The perforation is in the central, horizontal portion of the TFC and occurs in a more ulnar location than that seen with the traumatic injury that occurs in this region (class IA). Class IID lesions include TFC perforation in the central horizontal portion associated with lunate or ulnar chondromalacia and LT ligament perforation. Class IIE injuries include TFC perforation associated with lunate or ulnar chondromalacia, LT ligament perforation, and additional ulnocarpal arthritis.

Clinical diagnosis. Tears of the TFCC result in ulnar-sided wrist pain and tenderness often with a palpable or audible click when the forearm is rotated. Although some degenerative tears or defects may not be symptomatic [39,40], examination usually reveals point tenderness volar to the extensor carpi ulnaris tendon and just distal to the ulna. Plain radiographs may be completely normal or may be significant for positive ulnar variance with cystic changes of the ulnar aspect of the lunate consistent with ulnar impaction.

MR imaging and MR arthrography. The TFCC appears as a low-signal intensity on MR images. The TFC specifically is triangular on coronal sections with its apex attaching to the intermediate-signal intensity hyaline articular cartilage of the ulnar aspect of the sigmoid notch of the radius with separate superior and inferior bifurcate attachments [20,27]. On axial images, the TFC is shaped like an equilateral triangle with the apex converging on the ulnar styloid and the base attaching on the superior margin of the distal radial sigmoid notch. On sagittal images

it appears thicker on the volar and dorsal aspects. Its ulnar attachment may appear bifurcated with two bands of lower-signal intensity attached to the fovea at the base of the radial aspect of the ulnar styloid and to the ulnar styloid itself separated by a region of higher-signal intensity [20,27]. These sites of increased signal at the ulnar attachments and at the hyaline cartilage of the sigmoid notch of the radius should not be mistaken for detachments or tears. Both the distal and proximal surfaces of the TFC are depicted on MR images; information that is not available with wrist arthroscopy.

The prestyloid recess is an extension of the radiocarpal joint, which lies near the ulnar attachment of the TFC. Fluid collected in the prestyloid recess between the TFC and meniscus homologue produces increased signal intensity on T2-weighted images. Contrast may accumulate in this region normally at MR arthrography.

In older patients a signal may be seen within the low-signal TFC on T1-weighted and proton-density–weighted MR images that is thought to be caused by mucoid and myxoid degenerative changes. Degeneration of the TFC is frequently seen and often asymptomatic [39,40]. When there is degeneration of the TFC, MR imaging shows intermediate-signal intensity on short-echo-time images that does not increase on T2- or T2*-weighted images. If such signal changes do not communicate with the inferior or superior surface of the TFC, and if the signals are not any brighter on images with T2 contrast, then the TFC degeneration is not indicative of a tear. Degenerative changes may also manifest as thinning or attenuation of the TFC. Progressive degeneration of the proximal surface leads to erosion, thinning, and perforation of the TFC. In partial degenerative tears (Palmer classes IIA and B), the signal extends only to one articular surface. With complete tears the signal extends to proximal and distal articular surfaces. Discontinuity or fragmentation of the TFC may also be seen. Fluid collecting in the DRUJ is an important secondary sign, but the presence of fluid signal alone is not indicative of a TFC tear.

In MR imaging there are no specific differentiating features separating a traumatically-induced tear of the TFC from one caused by degeneration. The appearance of these lesions may also be similar in symptomatic and asymptomatic individuals; determining the clinical relevance of these lesions and their correlation with patients' symptoms may be difficult. Patient age, tear location, clinical history, and associated lesions are often criteria that may be needed to differentiate their origin.

Many studies have investigated the usefulness of conventional MR imaging in the detection of TFC tears [41–45]. Degenerative (Palmer class II) and traumatic radial tears of the TFC (Palmer classes IA and D) may be confidently appreciated with MR imaging. Reported sensitivities, specificities, and accuracies compared with arthroscopy are above 90% for degenerative (Palmer class II) tears [41,44,45]. Traumatic tears of the ulnolunate or ulnotriquetral ligaments (Palmer class IC) and traumatic avulsions of the TFCC from its ulnar insertion (Palmer class IB), however, are more difficult to diagnose. MR imaging findings that correlate with ulnar-sided type of tears include altered morphology of the ulnar attachments of the TFC, excessive fluid localized to this region, and linear fluid signal in the ulnar TFC itself extending to its surface. Oneson et al [44], in an MR imaging study of TFCC tears that underwent arthroscopic evaluation, had a poor sensitivity to ulnar tears for two observers (25% and 50%, respectively). Haims et al [41] most recently confirmed the limitations of conventional MR imaging in the diagnosis of peripheral ulnar-sided tears of the TFC. They found the evaluation sensitivity of tears of the peripheral TFCC to be 17%, with a specificity of 79% and an accuracy of 64%. The poor diagnostic accuracy was attributed to the presence of normal striated fascicles at the periphery of the TFCC and fluid that collected on the ulnar aspect of the wrist [41].

Not uncommonly, fluid signal and thickening may be present along the ulnar aspect of the TFCC. This appearance may be caused by degenerative or inflammatory changes, or as a result of a prior, healed peripheral TFC injury with scarring and chronic synovitis. This type of injury may be difficult to differentiate from an acute peripheral TFC injury. Correlation with the patient's clinical history is helpful to this end [45].

Although initial experience with MR arthrography in the assessment TFCC lesions did not show sufficient benefit over conventional MR imaging [3,10], the most recent MR arthrography studies using up-to-date technology have improved sensitivity, specificity, and accuracy for the evaluation of partial and complete tears of the TFCC [1,4,8]. In a recent MR arthrography study of 125 patients who had arthroscopic correlation, Schmitt et al [8] reported sensitivity of 97.1%, specificity of 96.4%, and accuracy of 96.8% for the detection of TFCC lesions.

The principal usefulness of MR arthrography resides in the evaluation of TFC peripheral ulnar tears. MR arthrography is also helpful in the detection of partial tears of the undersurface by revealing contrast extending through and outlining the TFC defect. Injection from the DRUJ may best outline such partial-thickness tears.

Treatment. The treatment of TFCC tears is complex and ongoing. Central tears (class IA) are treated with arthroscopic debridement by enlarging the tear in such a way that the fibrocartilage flaps can no longer make contact [46,47]. If a class IA tear is associated with an ulna-plus, simple debridement of the tear is insufficient and a formal ulnar shortening or an arthroscopic resection of the dome of the ulnar head (the so-called "arthroscopic wafer" procedure) is most appropriate [48,49]. The periphery of the TFCC is penetrate by blood vessels and has the ability to heal itself [31]. This has been opportunely used to reinsert (open surgery or an arthroscopic reinsertion) the peripheral type of TFCC tears (class IB) to the capsule with good results [50]. As for class IA, correction of a concomitant ulna-plus is mandatory: repair of a tear without considerations of the ulna discrepancy leads to failure of the procedure [51,52]. In time peripheral tears may lose the ability to heal themselves and repair is no longer possible. Whether because of long delays after the injury, or because no local tissues are available, reconstruction of DRUJ stability requires tendon graft plasty techniques [53]. Avulsion of the ulnocarpal ligaments (class IC) is rarely an isolated injury, but often part of a generalized ligamentous derangement [54]. Treatment depends on the stage and can be successfully performed by open surgery or arthroscopic treatment [47]. Avulsion of the TFC from the very edge of the radius (class ID) should be differentiated from the central-most type IAs because the latter are not repairable. In class ID, however, the TFC can be successfully reinserted into the radius and repaired with an open procedure or arthroscopic method [55].

Ligamentous anatomy

The carpal ligaments are divided into two major groups: in- and extrinsic ligaments [56]. The intrinsic, or intraosseous ligaments, are entirely within the carpus and connect the individual carpal bones. The extrinsic ligaments link the carpal bones to the radius and ulna. The description of the volar and dorsal extrinsic ligaments varies considerably [2,17,57–60], perhaps because of anatomic variation, but also that these ligaments are not discrete structures but rather focal thickenings of the fibrous capsule. The volar complex is the stronger of the two and seems to play a larger role in wrist stability. Recent anatomic studies suggest, however, that the dorsal ligaments of the wrist play an even larger and more important role in carpal stability and carpal kinematics than was previously recognized [61–63].

Volar ligaments
Volar radiocarpal ligaments.   The volar extrinsic radiocarpal ligaments are comprised of the radio-scaphocapitate, radiolunotriquetral or long radiolunate, radioscapholunate, and short radiolunate ligaments (Fig. 6) [2,17,56–58,60]. The radioscaphocapitate is a prominent ligament, which extends from the radial styloid through a groove in the waist of the scaphoid to the volar aspect of the capitate.

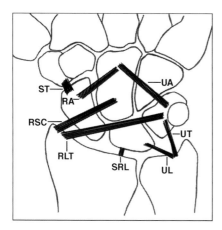

Fig. 6. Anatomy of the volar ligaments. Diagram illustrating the extrinsic radio- and ulnocarpal ligaments and intrinsic scaphotrapezial and deltoid ligaments. RA, radial arm of the deltoid ligament; RLT, radiolunotriquetral; RSC, radio-scaphocapitate; SRL, short radiolunate ligaments; ST, scaphotrapezial ligament; UA, ulnar arm of the deltoid ligament; UL, ulnolunate ligament; UT, ulnotriquetral ligament.

The radiolunotriquetral ligament arises from the volar aspect of the radial styloid. The radioluno-triquetral and radioscaphocapitate ligaments share the same radial attachment site, with the radiolunotri-quetral further extending to the ulnar side of the distal radial volar surface. The radiolunotriquetral ligament travels distally and ulnar-ward through the groove of the scaphoid proximal to the radioscaphocapitate ligament and widely inserts into the volar aspect of the triquetrum. The radiolunotriquetral ligament is made up of the radiolunate and LT portions. As such it has also been referred to as separate long radio-lunate and volar LT ligaments [17].

The radioscapholunate ligament, or ligament of Testut, arises from the volar aspect of the distal radius at the prominence between the scaphoid and lunate fossa, and extends distally inserting itself into the proximal volar aspect of the scaphoid, lunate, and SL ligaments. The radioscapholunate ligament, previ-ously thought to be an important scaphoid stabilizer, is now considered to be a neurovascular pedicle derived from the anterior interosseous and radial arteries and anterior interosseous (Fig. 7) [64]. The short radiolunate ligament, which is contiguous to the TFCC volar fibers, originates from the volar margin of the distal part of the radius and inserts into the proximal part of the volar surface of the lunate.

Volar ulnocarpal ligaments.   The volar ulnocarpal ligaments are comprised of the ulnolunate and ulno-triquetral ligaments. They originate at the volar edge of the TFC and insert into the lunate and the triquetrum, respectively [2,17,57–60]. The arcuate, or deltoid ligament, is a V-shaped volar intrinsic ligament with a capitotriquetral (ulnar) and a capito-scaphoid (radial) arm.

Dorsal ligaments
The dorsal radiocarpal ligament, also called dorsal radiotriquetral, arises from the dorsal aspect of the distal radius and extends distally over the dorsal aspect of the lunate to insert on the dorsal aspect of the triquetrum [62,63]. The dorsal radiocarpal has been grouped into four types by Viegas et al [63] according to a modification of Mizuseki and Ikuta's [62] classification: type I, the ligamentous fibers are attached to the dorsal margin of the distal ulnar aspect of the radius, then extend to the dorsal tubercle of the triquetrum (54%); type II, the same basic pattern as in type I with an additional ligamentous branch between the dorsal tubercle of the triquetrum and the dorsal margin of the distal radius at its extensor carpi radi-alis level (24%); type III, in addition to the type II pattern, there are more thin fibers spanning from the

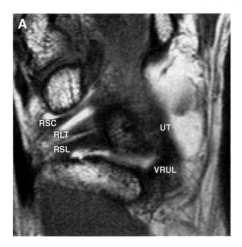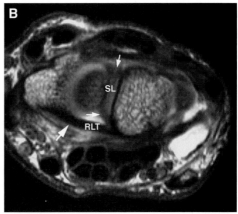

Fig. 7. Anatomy of the volar ligaments. (*A*) Coronal T1-weighted MR arthrogram image showing the radioscaphocapitate (RSC), the radiolunotriquetral (RLT), and the radioscapholunate (RSL) ligaments in the volar radial aspect of the wrist. The volar radioulnar (VRUL) and ulnotriquetral (UT) ligaments are seen in the ulnar aspect of the wrist. (*B*) Axial T1-weighted MR arthrogram image showing the thick radiolunotriquetral (RLT) ligament (*large arrow*). Note the dorsal and volar components (*small arrows*) of the scapholunate (SL) ligament.

dorsal triquetrum to the dorsal radius between the main ligament and the ligamentous branch (12%); and type IV, a type I pattern with additional separate ligamentous fibers from the ulnar aspect of the radius (9%) (Fig. 8) [63].

The dorsal intercarpal ligament originates from the dorsal tubercle of the triquetrum. It attaches to the dorsal distal aspect of the lunate and inserts into the dorsal groove of the scaphoid and dorsal proximal rim of the trapezium. The proximal side of this transversely oriented ligament is thicker than its distal side. This ligament attaches not only to the lunate and scaphoid but also to the SL ligament and the LT ligament.

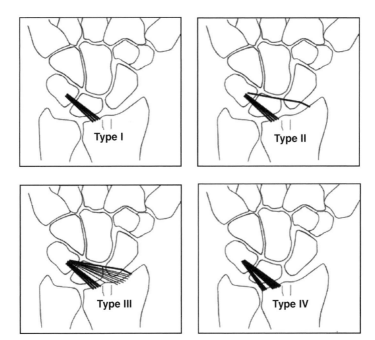

Fig. 8. Diagram illustrating the four types of dorsal radiocarpal ligaments.

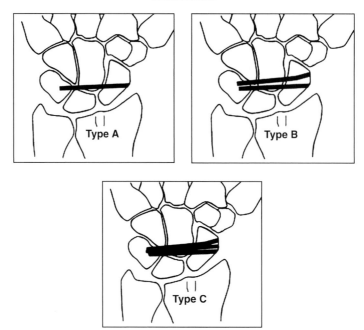

Fig. 9. Diagram illustrating the three types of dorsal intercarpal ligaments.

The dorsal intercarpal ligaments have been classified into three types: type A, a single thick fiber or net of thin fibers (30%); type B, two thick fibers (44%); and type C, three or more fibers (26%) (Fig. 9) [63].

Intrinsic ligaments

The most important intrinsic ligaments are the SL and LT ligaments (Fig. 10). These ligaments link the bones of the proximal carpal row on their proximal surfaces, separate the radio- from the midcarpal compartments, and provide flexible linkage so that the proximal carpal row functions properly. These ligaments have three components: (1) dorsal, (2) proximal, and (3) volar. Although the dorsal and volar components mostly consist of collagen fibers, the proximal, or membranous portion, is composed predominantly of fibrocartilage containing only a

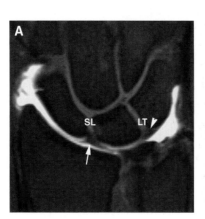

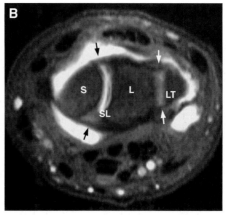

Fig. 10. Normal variations in the appearance of the scapholunate (SL) and lunotriquentral (LT) ligaments. (*A*) Coronal fat-suppressed T1-weighted MR arthrogram image showing the SL ligament with a triangular morphology (*arrow*) and a thick LT ligament with a linear morphology (*arrowhead*). (*B*) Axial fat-suppressed T1-weighted MR arthrogram image showing the dorsal and volar portions of the SL (*black arrows*) and LT ligaments (*white arrows*). L, lunate; S, scaphoid.

few superficial longitudinally-oriented collagen fibers [2,58].

Ligamentous injury

Intrinsic and extrinsic ligaments play an important role in wrist stability (Fig. 11). Injuries to the intrinsic ligaments are frequently associated with extrinsic volar and dorsal ligament lesions and often may be a cause of patients' chronic wrist pain and dysfunction. Wrist instabilities may be classified as static or dynamic [65]. Static instability refers to carpal malalignment that can be detected on standard posteroanterior and lateral radiographs, which indicate more significant and chronic ligamentous injuries [66]. Dynamic instability refers to carpal malalignment that is reproduced with physical examination maneuvers and when stress radiographs are made. With dynamic instability, there is no evidence of carpal malalignment on conventional radiographs [66].

Carpal instability is also classified into dissociative and nondissociative [59,67]. Dissociative carpal instability indicates an injury to one of the major intrinsic ligaments, such as that seen in SL dissociation and lunatotriquetral and perilunate dislocation, which may occur in association with injuries to the volar and dorsal extrinsic ligaments [66]. Nondissociative carpal instability refers to an injury to a major extrinsic ligament, with intact intrinsic ligaments, such as occurs in dorsal carpal subluxation, midcarpal instability, volar carpal subluxation, or capitate–lunate instability.

Mayfield [57] described a pattern of sequential four-stage ligamentous injury, called progressive perilunar, to be initiated on the radial aspect of the wrist and extending across the perilunate ligaments to the ulnar aspect of the wrist. Stage 1 consists of tearing the volar extrinsic radioscaphocapitate ligament with elongation or partial tearing of the SL ligament. In stage 2, with continued loading, total ligamentous failure occurs at the SL joint, followed by failure of the radioscaphocapitate ligament or an avulsion fracture of the radial styloid. Stage 3 refers to separation of the triquetrum from the lunate with associated injury to the radiolunotriquetral and dorsal radiocarpal ligament disruption. Finally, in stage 4 the ultimate failure of the dorsal radiocarpal ligament with volar lunate dislocation occurs.

Scapholunate instability

The most commonly injured ligament in the wrist is the SL [68]. SL injury usually occurs when the wrist is in hyperextension and ulnar deviation during axial compression and carpal supination. Anatomic and biomechanical testing has shown that the dorsal portion of this ligament is thicker and stronger than its volar and proximal portions [69]. It seems, however, that both the dorsal and volar components have important roles in normal SL stability. The scaphoid attachment of the SL ligament is more likely to avulse than in the stronger lunate attachment. In fact, an SL ligament tear may be associated with a scaphoid avulsion fracture.

Disruption of the SL ligament is a prerequisite for the development of SL dissociation (Fig. 12) [66]. Although isolated SL disruption does not cause an immediate and complete diastasis or abnormal radiographic alignment, biomechanical testing has shown

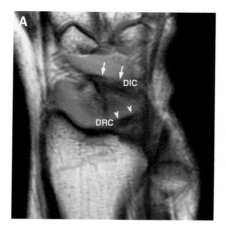
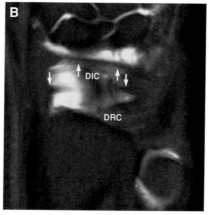

Fig. 11. Normal anatomy variants of the dorsal extrinsic ligaments. (*A*) Coronal T1-weighted MR arthrogram image showing the most common configuration of the dorsal extrinsic ligaments, the dorsal radiocarpal (DRC) ligament type I (*arrowheads*) and the dorsal intercarpal (DIC) ligament type B (*arrows*). (*B*) Coronal fat-suppressed T1-weighted MR arthrogram image showing DRC type III and DIC type C (*arrows*).

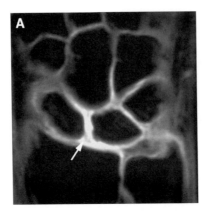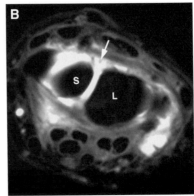

Fig. 12. Scapholunate ligament tear. (*A*) Coronal fat-suppressed T1-weighted MR arthrogram image revealing a complete SL ligament tear with a slight widening of the SL interval (*arrow*). (*B*) Axial fat-suppressed T1-weighted MR arthrogram image showing a complete tear of the dorsal component of the SL ligament (*arrow*). L, lunate; S, scaphoid.

that loss of this critical structure results in significant changes in contact load patterns and kinematics. Because the scaphoid and lunate have other strong extrinsic supporting ligaments, one or more of these ligaments must also have sustained damage to show significant radiographic changes. Presently, it is recognized that involvement of dorsal intercarpal land volar scaphotrapezial ligaments needs to occur to complete SL dissociation (Fig. 13) [62,63,70]. When left untreated, disruption of this ligament leads to progressive flexion posture of the scaphoid (rotary subluxation) and migration of the scaphoid away from the lunate. With time, a pattern of degenerative arthritis develops. Watson and Ballet [71] coined this pattern "SL advanced collapse," and postulated that it is the most common pattern of degenerative wrist

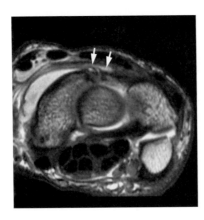

Fig. 13. Dorsal intercarpal ligament tear in a patient who has a dorsal wrist sprain. Axial T1-weighted MR arthrogram image showing thickening and fraying of the dorsal intercarpal ligament near the scaphoid attachment (*arrows*).

arthritis. Stage I describes isolated degenerative changes between the radial styloid and distal scaphoid. Stage II describes progression of the arthrosis to the proximal scaphoid fossa and proximal scaphoid. Stage III heralds involvement of the midcarpal joint at the capitate-lunate articulation. Finally, stage IV represents the development of pancarpal degenerative arthritis.

Patients who have SL ligament injuries experience pain with direct palpation over the SL interval. Wrist motion is only appreciably impaired after the development of carpal degenerative arthritis [66,68].

Plain radiographs are recommended in the initial evaluation of suspected SL injury. In a patient with static SL instability, a neutral posteroanterior radiograph reveals an increase in the SL interval. The clenched fist view accentuates the spacing between the scaphoid and lunate [66,68].

Instabilities involving the SL articulation may also produce a dorsal intercalated segmental instability pattern, which refers to the appearance of the lunate—the intercalated segment—on the lateral radiograph. In this pattern of instability, the lunate is dorsally angulated in the sagittal plane and the capitate is dorsally displaced toward the radiometacarpal axis (radiolunate angle, > 10° degrees). The angle formed between the longitudinal axes of the scaphoid and lunate, which normally measures 45° (±15°), is increased and greater than 70° [66,68].

MR imaging and MR arthrography. The SL ligament displays a triangular (90%) or linear morphology (10%) [23]. In 63% of cases studied by Smith [23,25], the SL ligament was seen as a homogeneous low or low-to-intermediate signal intensity structure. In 37% of cases, there were intermediate-signal intensity areas traversing portions of the SL ligament, which can be potentially mistaken for tears.

Partial tears and elongated, but intact, ligaments may be visualized with MR imaging. A partial tear may be diagnosed when there is focal thinning or irregularity or high-signal intensity in a portion of the ligament, more commonly occurring in the central and volar portion where the weakest ligamentous attachments are located. Complete tears of the SL ligament appear as distinct areas of discontinuity within the ligament, or an absence, altogether, of the ligament. Fluid in the midcarpal joint is a sensitive, but nonspecific finding of ligament tears [20]. In more advanced cases, widening of the SL ligament articulation may be evident if complete SL ligament tears are associated with involvement of dorsal intercarpal ligament and volar scaphotrapezial ligaments [62,63,70].

MR arthrography can potentially evaluate the precise location and exact magnitude of any ligamentous defect and differentiate those lesions that may involve only the central membranous portion and be of degenerative origin [6,7]. These central lesions may be painful but not indicative of instability as would involvement of the other ligaments, particularly the dorsal portion. MR arthrography may yield an increased sensitivity to SL tears over MR imaging, especially in more subtle injuries. This includes partial tears, which may show contrast leak or imbibition into a portion of an injured ligament or better outline morphologic alterations or stretching. MR arthrography may also help outline dysfunctional ligaments that may have healed over with fibrosis or are scarred and increases the accuracy of peripheral ligament avulsions where the ligament has not lost its normal morphology. The latter may be evident clinically, but difficult to document with conventional MR imaging. In complete tears, MR arthrography shows contrast material communication between the radio- and midcarpal compartments [6,7].

The diagnostic value of MR imaging for detecting SL ligament lesions is controversial. Conventional MR imaging sensitivity ranges from 50% to 93%, specificity from 86% to 100%, and accuracy from 77% to 87% as compared with arthroscopy and surgery [6,7,20].

A recent study by Schmitt et al [8] in 125 patients suffering from wrist pain who were examined with double MR arthrography with arthroscopic correlation revealed a sensitivity of 91.7%, specificity of 100%, and accuracy of 99.2% for the detection of complete tears of the SL ligament, and a sensitivity of 62.5%, specificity of 100%, and accuracy of 95.2% for partial tears.

Treatment. Acute injuries with complete SL ligament rupture and overt dissociation should be treated as early as possible because ability of the area to heal diminishes rapidly. Often repairs are impossible after a period of 6 weeks, let alone 3 months. The ligament is usually avulsed from the scaphoid. Its reinsertion and repair is performed with transosseous stitching or mini bone anchors [72]. Nondestabilizing acute SL injuries respond to simple immobilization for 4 to 6 weeks, if minor, but require percutaneous Kirschner wiring if the injury is moderate to severe. Treatment of isolated lesions to the most proximal part of the ligament—the membranous portion—requires only arthroscopic debridement because it is mechanically inconsequential. Shaving permits immediate mobilization and rapid recovery [72].

A myriad of surgical management techniques have been performed on chronic injuries in the 1980s and 1990s, from SL arthrodesis to different ligament reconstructions [73]. Reconstruction of all structures involved (SL, dorsal intercarpal ligament, and volar scaphotrapezial ligaments) seems to be the most logical way of dealing with complete SL dissociation [74].

Lunotriquetral instability

LT injuries occur approximately one sixth as commonly as SL injuries [75,76]. Tears of the LT ligament may coexist with static and dynamic patterns of volar midcarpal instability. In patients who have volar midcarpal instability, the lunate is no longer linked to the triquetrum and follows the scaphoid. In this situation the lunate angulates palmarly (radiolunate angle, 10 degrees in a volar direction), which causes the capitate to become displaced palmar to the radiometacarpal axis. The angle formed between the longitudinal axes of the scaphoid and lunate is less than 30 degrees. An anatomic and biomechanic study performed by Viegas et al [63] found that the dorsal radiocarpal ligament must be attenuated or disrupted for a static volar midcarpal instability to develop.

Patients who have LT instability present with ulnar-sided pain. The examination of a patient who has an LT ligament tear typically reveals point tenderness over the LT interval. Plain radiographs of the wrist are usually normal in isolated LT tears. As progressive injury occurs to secondary constraints, the wrist assumes a volar midcarpal instability configuration consistent with midcarpal instability (Fig. 14) [75,76].

MR imaging and MR arthrography. The LT ligament is consistently visible with MR imaging when

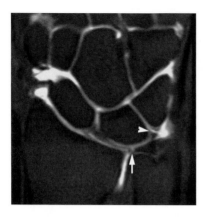

Fig. 14. Lunotriquetral ligament tear. Coronal fat-suppressed T1-weighted MR arthrogram image showing avulsion of the ulnar aspect of a delta-shaped LT ligament (*arrowhead*). Note the traumatic radial tear of the TFC (*arrow*) (Palmer class IA).

it is present. This ligament is triangular in 63% and linear in 37% of patients [24]. The signal intensity is homogeneous and low in 75%, but a brighter, linear signal traverses part or all of the LT ligament in 25% and is distinguishable from a tear because it is not quite as bright as a fluid signal.

Disruption of the ligament is shown on T2*- or fat-suppressed T2-weighted images as either complete ligamentous disruption or a discrete area of bright, linear signal intensity in a partial or complete tear. Absence of the ligament is not as useful a finding because the LT ligament may not be as reliably observed on MR imaging as the SL ligament [4,10,20,25]. Small membranous perforations may exist in the presence of intact dorsal and volar portions of the LT ligament. In fact, most degenerative perforations occur in the thin, membranous portion of the LT ligament and are difficult to appreciate on MR images. Osseous widening of the LT articulation is not usually evident even in advanced cases. Of note, the volar portion of the ligament attaches to the TFC here, which results in a discontinuous appearance—a diagnosing pitfall to be avoided.

Conventional MR imaging is less accurate in LT than SL ligament evaluation [10,11,20,22,25]. A wide range of sensitivities (40%–100%) and specificities (33%–100%) have been reported in the diagnosis of LT ligament perforations using MR imaging [4,10,20,25,26]. MR arthrography is superior to conventional MR imaging, in that it allows for identification of the size, morphology, and location of an LT ligament tear. This information is crucial because communication through a pinhole, small

perforation, or deficiency in the thin membranous portion of the ligament may be insignificant in the presence of grossly intact dorsal and volar ligaments. In complete tears, contrast material communication between the radio- and midcarpal compartments may be identified [1,4,7,8].

Treatment. LT and LS ligament injury management guidelines and aftercare are similar: if diagnosed early (no later than 3 weeks after trauma) primary repair is possible and performed with transosseus reinsertion of the ligament. Chronic injuries are best currently managed through reconstruction of the ligament instead of arthrodesis, which carries a worse prognosis and high nonunion rate [77]. A tendon graft, consisting of half the extensor carpi ulnaris, is divided proximally and left distally attached, then passed through drill holes in the triquetrum and lunate, and finally sutured back to itself.

Extrinsic carpal ligaments injuries

Tears of the extrinsic volar or dorsal carpal ligaments are not commonly identified with MR imaging. Dorsal ligaments provide stability in wrist motion and are frequently injured when falling on the outstretched hand and result in a sprain of the dorsal wrist [22].

MR imaging can visualize tears of these ligaments, which appear with increased signal intensity, irregularity, and fraying, and correspond to high signal intensity on T2 sequences. MR arthrography could potentially evaluate dorsal and volar extrinsic ligaments tears because of its high resolution and excellent contrast between ligaments and surrounding structures; however, its exact use in the evaluation of these lesions is yet to be clearly defined [2,17,20,22].

Midcarpal instability

Injuries where the ligaments between the proximal carpal row bones remain intact, but the proximal and distal rows unstable is not very common [78]. This type of injury has been classified as nondissociative carpal instability [67]; an abnormality that is thought to be the result of tears, laxity, or insufficiency of the deltoid ligament ulnar limb [20,67,78]. There is also evidence of generalized ligamentous laxity. With ulnar deviation of the wrist there is a so-called "catch-up clunk" where a sudden painful snap of the proximal row occurs as the wrist is ulnarly deviated [78].

MR imaging and arthrography can visualize this kind of injury. MR arthrography allows a more precise evaluation of the ligament, although its accuracy in depicting its disruption in patients who have clinically evident midcarpal instability has yet to be reported.

Ulnar wrist pain

Ulnar wrist pain has often been equated with low back pain because of its insidious onset, vague and chronic nature, intermittent symptoms, and the frustration that it induces in patients. Ulnar wrist pain frequently may be caused by a broad spectrum of osseous or soft tissue disorders including TFCC tears, DRUJ arthritis and instability, LT ligament disruption, Kienböck's disease, pisotriquetral arthritis, extensor carpi ulnaris lesions, and/or ulnar-sided wrist impaction syndromes [75]. The latter syndromes constitute a group of pathologic entities that result from repetitive or acute forced impaction between the distal ulna and ulnar carpus or distal radius and surrounding soft tissues and results in bone or soft tissue lesions [79,80].

In an adequate clinical setting, conventional radiographic findings of anatomic variants or pathologic conditions of the ulnar wrist can suggest the diagnosis of a given ulnar-sided impaction syndrome. Often diagnosis is difficult or delayed, however, because symptoms and clinical findings are usually nonspecific and similar among the different pathologic conditions in the ulnar-sided wrist. Moreover, significant disease and incapacitating pain may be present despite minimal evidence from conventional radiography. Conventional MR imaging and arthrography allow earlier detection of the bone and soft tissue lesions that are present in the different ulnar-sided wrist impaction syndromes [79,80].

Ulnar impaction syndrome

Ulnar impaction syndrome, also known as ulnar abutment or ulnocarpal impaction, is a degenerative condition characterized by chronic impaction between the ulnar head and the TFCC and ulnar carpus and results in a continuum of pathologic changes: TFC degenerative tear; chondromalacia of the lunate, triquetral, and distal ulnar head; LT ligament instability or tear; and, finally, osteoarthritis of the DRUJ and ulnocarpal joint.

The pathologic changes appearing in ulnar impaction syndrome most commonly occur with positive ulnar variance but can occasionally occur with neutral or negative ulnar variance [81]. The most common predisposing factors include congenital positive ulnar variance, malunion of the distal radius, premature physeal closure of the distal radius, Essex–Lopresti fracture, and previous surgical resection of the radial

head. All of these predisposing factors result in a fixed increase in underlying ulnar loading associated with relative lengthening of the ulna or increased dorsal tilt of the distal radius [79,80].

In the absence of obvious structural abnormalities, ulnar impaction syndrome may result from daily activities that cause excessive intermittent loading of the ulnar carpus. It has also been shown that asymptomatic changes in ulnar impaction syndrome develop over time, so that this condition may be present even if symptoms are not evident.

The clinical manifestation of ulnar impaction syndrome generally consists of chronic or subacute ulnar wrist pain, often exacerbated by activity and relieved by rest. Pronation grip radiographs are useful in determining the increase in ulnar variance and the impaction between ulnar carpus and the dome of the ulnar head [38]. Underlying abnormalities, including malunion of a distal radial fracture with residual radial shortening and abnormal dorsal tilt, may be present. Secondary changes in the ulnar carpus include subchondral sclerosis and cystic changes in the ulnar head, ulnar aspect of the proximal lunate, and proximal radial aspect of the triquetral [82]. MR imaging and MR arthrography are helpful in detecting radiologically occult lesions [79,80,82]. Fibrillation, or partial-thickness defects of articular cartilage, in the ulnar wrist can be detected by MR arthrography; however, the accuracy of MR arthrography tends to be more favorable for high-grade or large cartilage defects [8,79,80]. Bone marrow edema can even be seen in patients who do not have cartilaginous degeneration revealed at arthroscopy, indicating that it is also a sensitive sign of ulnar impaction. Progression of the syndrome results in sclerotic changes, which appear as areas of low signal intensity on both T1- and T2-weighted images and subchondral cysts, which appear as well-defined areas of low signal intensity on T1- and high signal intensity on T2-weighted images [79,80,82].

MR arthrography may be necessary in selected cases to clarify the exact stage of ulnar impaction in the preoperative evaluation. This technique is especially useful in determining any perforation of the TFC (Palmer class IIB versus IIC) and in determining the status of the ulnocarpal ligaments and the LT ligament (Palmer class IIC versus IID) (Figs. 15 and 16).

Briefly, there are three types of procedures that can be used when dealing with symptomatic ulnar impaction: (1) the wafer procedure [48], (2) the arthroscopic wafer procedure [49], and (3) the formal ulnar shortening procedure [51]. All three aim to recede the dome of the head of the ulna. In the wafer

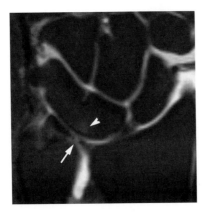

Fig. 15. Ulnar impaction syndrome in a patient who has LT coalition Minaar's type III and positive ulnar variance (Palmer class IIC lesion). Coronal fat-suppressed proton-density-weighted spin-echo MR arthrogram image showing degenerative central perforation of the TFC (*arrow*), chondromalacia, and subtle bone marrow edema in the ulnar aspect of the lunate bone (*arrowhead*).

procedure, the distal-most 2 to 3 mm of the dome of the ulnar head are resected by a limited approach centered on the DRUJ; this can be performed by arthroscopic instrumentation (arthroscopic wafer procedure) [48,83]. In the ulnar shortening procedure, the distal shaft of the ulna is approached dorsolaterally and the desired amount of ulna sliced away, followed by rigid fixation. Choosing which procedure to perform depends on several considerations: the amount of ulnar variance, Palmer type, shape of the sigmoid fossa and ulnar seat, presence of concomitant LT instability, and skill level of the surgeon [79,80].

Ulnar styloid impaction syndrome

Ulnar styloid impaction syndrome refers to a group of pathologic entities characterized by impaction between the ulnar styloid and the triquetrum and the surrounding soft tissues [80,84,85]. Garcia-Elias [86] has developed the ulnar styloid process index, a method that assesses the relative size of the ulnar styloid. An excessively long ulnar styloid has an ulnar styloid process index greater than 0.21 ± 0.07 or an overall length greater than 6 mm.

An elongated ulnar styloid is the most common variant implicated in the development of ulnar styloid impaction syndrome. Another anatomic variant is ulnar styloid process volarly or radially curved with a parrot-beaked appearance, which reduces significantly the styloid-carpal distance (Fig. 17) [80].

An enlarged ulnar styloid can be an anatomic variant or secondary to the malunion of the avulsion fracture at the fovea of the ulna. This ulnar styloid morphologic variation reduces the ulnar joint space causing repetitive impaction between ulnar styloid and ulnar aspect of the lunate bone and radial aspect of the triquetral [80].

Two types of ulnar styloid nonunion have been classically described anatomically and their different treatments discussed (Fig. 18) [87]. Type 1 is defined as a nonunion associated with a stable DRUJ and affects only the tip of the styloid, while the TFCC remains intact because its major attachments are at

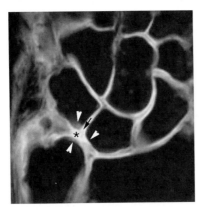

Fig. 16. Ulnar impaction syndrome (Palmer class IID lesion). Coronal fat-suppressed T1-weighted MR arthrogram image showing chondromalacia of the ulnar head, ulnar side of the lunate bone, and radial side of the triquetral bone (*arrowheads*); central perforation of the TFC (*asterisk*); and LT ligament tear (*arrow*).

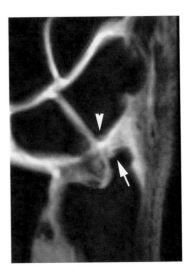

Fig. 17. Ulnar styloid impaction syndrome. Coronal fat-suppressed T1-weighted MR arthrogram image showing an excessively long ulnar styloid process, chondromalacia of the triquetral bone (*arrowhead*) and ulnar styloid tip (*arrow*), and LT ligament perforation.

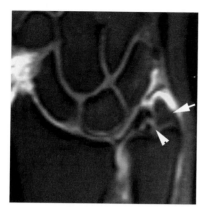

Fig. 18. Ulnar styloid (US) impaction syndrome secondary to type II nonunion of the US process. Coronal fat-suppressed T1-weighted MR arthrogram image revealing nonunion of the US process (*arrow*) associated with ulnar avulsion of the TFC complex (Palmer class IB TFC complex injury) (*arrowhead*).

the base of the styloid. Type 2 is defined as a non-union associated with DRUJ subluxation and is the result of an avulsion of the ulnar attachment of the TFCC (Palmer class IB lesion). The diagnosis of ulnar styloid impaction syndrome is based primarily on the clinical history and physical examination of the patient, and supported by radiographic evidence of morphologic variations or pathologic conditions of the ulnar styloid [80,84,85].

MR arthrography is an excellent modality for visualizing the integrity of the TFCC and its ulnar attachments, presence of nonunited bone fragments, and associated chondromalacia and subchondral bone changes of the carpus. The presence of synovitis and joint effusion in the ulnocarpal joint is also common [80].

Resection of all but the 2 proximal-most mm of the ulnar styloid (avoiding interference with TFCC insertion) is the treatment of choice in the ulnar styloid impaction syndrome secondary to elongated or parrot-beaked ulnar styloid [79,80]. Surgical treatment of the ulnar styloid impaction syndrome secondary to ulnar styloid nonunion should start with diagnostic arthroscopy that allows for classification of the condition under one of the aforementioned subtypes. If the TFCC is lax showing a positive "trampoline sign," the ulnar styloid (and with it, the TFCC) should be reinserted into the fovea and appropriately fixed by a limited incision in the ulnar aspect of the wrist. If no TFCC loosening is appreciated, the offending bony fragment should be removed [79,80].

Hamatolunate impaction syndrome

Hamatolunate impaction is a rare cause of ulnar-sided wrist pain secondary to chondromalacia of the proximal pole of the hamate bone in patients who have lunate bones with a medial articular facet on the distal surface that articulates with the hamate bone (type II lunate) [88–90]. It has been suggested that the repeated impingement and abrasion of these two bones is the mechanism that develops into chondro-malacia when the wrist is in use in the full ulnar deviation position. Patients experience pain in full ulnar deviation of the wrist especially when combined with holding the distal carpal row in forced supination first.

MR arthrography may reveal defects of the articular cartilage, bone edema, sclerosis, and subchondral cysts in the proximal pole of the hamate bone [80,89]. Arthroscopic burring of the apex of the hamate represents state-of-the-art treatment of this condition [79,80].

Postoperative wrist

The increased number of patients who undergo arthroscopy and open surgical techniques to repair internal derangements of the wrist has produced an increasing demand for postoperative MR imaging evaluation because of poor outcome, recurrent symptoms, or new injury [12–14]. Several factors, however, may decrease the accuracy of MR imaging in the assessment of the postoperative wrist. These factors include surgical distortion of native anatomy, changes in signal intensity of the tissues, and image degradation caused by metallic artifacts. Radiologists should be familiar with the more common procedures currently used to repair injuries of the wrist and typical MR imaging findings in each postoperative situation to be able to recognize complications associated with such procedures [12–14].

Conventional MR imaging of postoperative TFC has unreliable results. Conventional diagnostic criteria used to diagnose TFC tears cannot be applied to postoperative TFC. Once injured, the TFC may never return to its normal, preinjury-signal intensity. Furthermore, an area of TFC healing with granulation tissue or fibrosis may appear as an abnormal signal reaching the articular surface and subject to be misinterpreted as a new tear. Additionally, TFC morphology following arthroscopy repair is abnormal, and this distortion and shape irregularity may be interpreted as a TFC tear. Specific signs of a TFC retear are a fluid-like signal within the TFC on T2-weighted images or a displaced TFC fragment [12–14]. To improve the TFC retear diagnostic accuracy, it is important that the preoperative images or operative reports are available for correlation with postoperative images.

MR arthrography has several advantages compared with conventional MR imaging in the evaluation of the postoperative TFC. MR arthrography uses T1-weighted images, which have a higher signal-to-noise ratio and frequently greater spatial resolution than T2-weighted images. Retears are diagnosed on MR arthrography when gadolinium signal intensity is seen extending into the TFC. In addition, increased intra-articular pressure allows for distention of normally apposed structures, such as the edges of a nondisplaced TFC tear.

MR arthrography is particularly valuable in the postoperative evaluation of symptomatic patients who have undergone reinsertion of the TFCC at the ulnaor radius (classes IB and D, respectively). TFC signal alterations extending to the TFC surface and abnormal TFC morphology can be seen at the reattachment sites on conventional MR imaging. MR arthrography with radiocarpal and DRUJ injection aids in the detection of focal TFC fillings or communicating defects in these patients, indicative of retear or inadequate reinsertion. In patients who have undergone TFC debridement or arthroscopic wafer procedure, MR arthrography is useful in identifying new TFC tears and chondral or osseous abnormalities that may occur after TFC repair (Fig. 19). MR arthrography can allow for a better evaluation of ligamentous repair techniques. MR arthrography findings become less valuable, however, when surgical reports are not available.

Ulnar collateral ligament injury

Injury to the UCL is a frequent lesion that occurs from a radially directed force on the abducted thumb [20,91,92]. This lesion was coined originally from Scottish gamekeepers who developed a chronic ligamentous strain or "gamekeeper's thumb" induced by repetitive stress from the method used to kill rabbits. This type of injury is now most commonly associated with skiing and know as "skier's thumb." Rupture of the UCL may be total or partial and usually takes place at its phalangeal point of insertion, but the rupture can also appear at its metacarpal insertion or in its midsubstance. It may be accompanied by an avulsion fracture of the proximal phalanx [20,91,92]. The rupture of the thumb UCL can be an isolated lesion or occur in combination with other joint structures, such as the volar plate or dorsal capsule. When the ligament ruptures distally, retraction may be associated with the interposition of the adductor pollicis aponeurosis with the torn UCL lying superficially at the proximal end of the aponeurosis. This injury, called a Stener lesion, can inhibit proper healing of the ligament [20,91,92].

At physical examination, a complete UCL tear induces the appearance of a palpable mass in the ulnar aspect of the joint and instability to radial stress reaching an angle of 30° or higher when compared with the contralateral thumb. Nevertheless, the differentiation between a nondisplaced UCL tear and a Stener lesion may be difficult in the acute setting because of overlying soft tissue edema and hematoma [20].

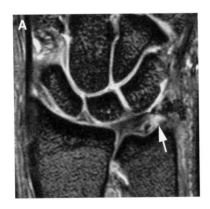
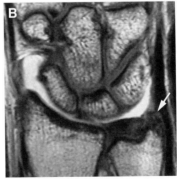

Fig. 19. Postoperative wrist. Patient who has a history of arthroscopic reinsertion of the TFC ulnar attachment (Palmer class IB lesion). (*A*) Coronal T2*-weighted conventional MR image showing persistent high-signal intensity at the TFC ulnar attachment (*arrow*) and small metallic artifacts in the ulnar aspect of the wrist. No confidence exists about the integrity of TFC ulnar attachment. (*B*) Coronal T1-weighted radiocarpal MR arthrogram image showing integrity of TFC ulnar insertion (*arrow*) with no communication between radiocarpal and DRUJ compartments.

MR imaging can detect the torn ligament and reveal displacement if present. MR imaging may also reveal any associated clinically-occult injuries involving bone or soft tissues. Primary signs of acute ligament tear include discontinuity, detachment, or thickening of the ligament associated with increased intraligamentous signal intensity on T2-weighted images. Secondary signs of acute ligament injury include soft tissue edema or hemorrhage, joint effusion, and bone bruises. MR imaging findings of a Stener lesion include UCL disruption from the base of the proximal phalanx with retraction or folding of the ligament. The ligament usually appears as a rounded or stumplike area of low-signal intensity lying superficially to the adductor aponeurosis. This characteristic MR imaging appearance has been described as a "yo-yo on a string" [20,92].

MR arthrography is more sensitive and accurate that MR imaging in the evaluation of acute and chronic UCL tears (Figs. 20 and 21) [91]. The joint distention obtained with MR arthrography allows precise assessment of the thickness of the ligament and its integrity at the insertion site. Acute or sub-acute tears of the UCL result in extravasation of contrast material into the adjacent soft tissues. Although MR arthrography is accurate at determining the presence of UCL ruptures and Stener lesions, the real value of MR arthrography lies in the detection of chronic ligament lesions. In this latter case, patients present with chronic pain and often joint instability. In chronic tears, secondary signs disappear and the ligament can show thickening, elongation, and an irregular or wavy contour [91].

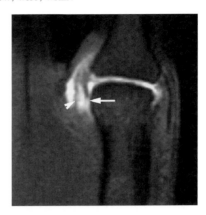

Fig. 21. Acute ligamentous strain of the UCL MCP thumb joint. Coronal fat-suppressed T1-weighted MR arthrogram image showing ligament rupture at its metacarpal insertion (*arrow*) with extravasation of contrast material into the adjacent soft tissues (*arrowhead*).

Nondisplaced UCL tears are usually treated conservatively. Surgical intervention is usually reserved for Stener lesions and complete undisplaced tears with laterolateral instability because conservative treatment leads to chronic instability and arthrosis [93,94]. Avulsion fractures involving more than 20% of the articular surface may require pinning.

Trapeziometacarpal instability

Mechanical instability of the trapeziometacarpal joint of the thumb is an important factor that may lead to articular degeneration of the joint and interfere with the normal function of the hand [95]. The anterior oblique ligament, also known as the "beak" or volar ligament, is an important joint stabilizer that limits its physiologic and radiodorsal subluxation. This ligament is a thick and broad structure that originates from the palmar tubercle of the trapezium and inserts into the beak at the base of the first metacarpal. The anterior oblique ligament is prone to rupture from a fall onto an outstretched hand where the point of contact is the base of a supinated and extended thumb metacarpal. Rupture of this ligament may cause symptomatic instability and an increased pressure on the incongruous articular cartilage, which may lead to osteoarthritis [95].

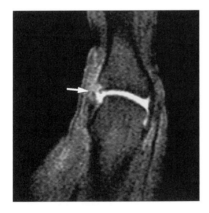

Fig. 20. Chronic ligamentous strain of the UCL MCP joint of the thumb in a patient who has chronic pain and joint instability. Coronal fat-suppressed T1-weighted MR arthrogram image showing chronic UCL ligament thickening and tear and the irregular contour at its phalangeal point of insertion (*arrow*).

MR arthrography is useful in the detection of acute or chronic ruptures of the anterior oblique ligament. In acute ruptures, MR arthrography reveals stretching or discontinuity of the ligament with contrast extravasation. The most frequent findings in chronic rupture are elongation, thickening, and an irregular and wavy contour (Fig. 22).

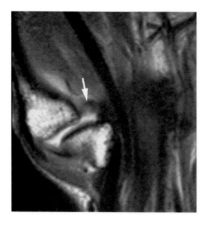

Fig. 22. Trapeziometacarpal instability. Coronal T1-weighted MR arthrogram image revealing a chronic rupture of the anterior oblique ligament with elongation, thickening, and a wavy contour of the ligament (*arrow*).

In most patients conservative treatment gives good results. Patients who have symptomatic instability may require reconstruction of the ligament to avoid any compromise to the function of the thumb [95].

Metacarpophalangeal lesions of the fingers

Collateral ligament injuries in the metacarpophalangeal joints of the fingers are rare compared with those of the thumb. Almost all collateral ligament tears in the fingers occur on the radial side [96,97]. The most commonly involved digit is the index finger, followed by the little finger. On the ulnar side of the little finger, displacement of the torn ligament over the intact sagittal band of the extensor hood may occur, similar to the Stener lesion of the first metacarpophalangeal joint. The displacement of the torn ligament interferes with healing and necessitates surgery [97].

MR arthrography can be used in the evaluation of selected cases of collateral ligament tears (ligamentous tears with articular instability) and volar plate lesions of the metacarpophalangeal joints. MR arthrography imaging criteria for diagnosis of acute collateral ligament tears include discontinuity, detachment, or thickening of the ligament. Extravasation of contrast solution into the adjacent soft tissues may also be observed in acute or subacute settings. Chronic tears often show thickening of the ligament, which is probably secondary to scar formation. Elongation, or a wavy ligament contour ligament, may also be seen. Injuries of the volar plate can be seen in MR arthrography as a disrupted attachment with a gap and ventral extravasation of contrast where avulsion of the volar plate takes place.

Although most lesions are treated conservatively, surgery has been advocated in cases of severe metacarpophalangeal instability or intra-articular displacement of the torn ligament [97].

Indirect MR arthrography

Indirect MR arthrography is based on the premise that intravenous administration of gadolinium followed by 5 to 10 minutes of light exercise leads to an enhancement effect in joints [98–100]. Indirect MR arthrography has been developed as an alternative, less invasive imaging technique than direct MR arthrography for joint assessment. Homogeneous enhancement of joint structures can be achieved with indirect MR arthrography, but it lacks controlled capsular distention. Some authors have reported good initial results with this technique in the evaluation of the TFCC and intrinsic ligaments of the wrist; however, a recent study concluded that although indirect MR arthrography improves sensitivity in the evaluation of the SL ligament when compared with conventional MR imaging, it does not significantly improve the ability to evaluate the TFCC or the LT ligament [99]. The major advantage of indirect arthrography of the wrist lies in detection of abnormalities other than the usual internal derangements. With indirect MR arthrography there is enhancement of fluid within the tendon sheath in the presence of tenosynovitis. Enhancement around flexor tendons even without increased fluid is a relatively specific sign of carpal tunnel syndrome. Subtle cartilage defects can be identified by associated subchondral enhancement [98].

Summary

Radiocarpal MR arthrography is an excellent technique for determining the localization, size, and extent of pathologic lesions of the TFCC and the intrinsic ligaments of the proximal row of the wrist. Triple-compartment MR arthrography is useful in the evaluation of patients who have refractory pain or instability syndromes of the wrist. Other evolving applications of MR arthrography of the wrist are ulnar-sided impaction syndromes and postoperative evaluation of TFCC and ligaments.

MR arthrography also may be applied in the diagnosis of selected pathologies of the small joints of the hand, such as UCL tears of the metacarpopha-langeal of the thumb, tears of the anterior-oblique ligament of the trapeziometacarpal joint, and lesions of the collateral ligaments and volar plate tears of the metacarpophalangeal joints of the fingers. MR arthrography cannot replace wrist arthroscopy; how-ever, it can facilitate the diagnosis and the indication for surgery and reduce the number of diagnostic arthroscopic interventions.

References

[1] Braun H, Kenn W, Schneider S, et al. Direct MR arthrography of the wrist: value in detecting complete and partial defects of intrinsic ligaments and the TFCC in comparison with arthroscopy. Rofo Fortschr Geb Rontgenstr Neuen Bildgeb Verfahr 2003;175: 1515–24.

[2] Brown RR, Fliszar E, Cotten A, et al. Extrinsic and intrinsic ligaments of the wrist: normal and pathologic anatomy at MR arthrography with three-compartment enhancement. Radiographics 1998;18:667–74.

[3] Carrino JA, Smith DK, Schweitzer ME. MR arthro-graphy of the elbow and wrist. Semin Musculoskelet Radiol 1998;2:397–414.

[4] Kovanlikaya I, Camli D, Cakmakci H, et al. Diagnostic value of MR arthrography in detection of intrinsic carpal ligament lesions: use of cine-MR arthrography as a new approach. Eur Radiol 1997;7: 1441–5.

[5] Palmer WE. MR arthrography: is it worthwhile? Top Magn Reson Imaging 1996;8:24–43.

[6] Scheck RJ, Kubitzek C, Hierner R, et al. The scapholunate interosseous ligament in MR arthro-graphy of the wrist: correlation with non-enhanced MRI and wrist arthroscopy. Skeletal Radiol 1997;26: 263–71.

[7] Scheck RJ, Romagnolo A, Hierner R, et al. The carpal ligaments in MR arthrography of the wrist: correlation with standard MRI and wrist arthroscopy. J Magn Reson Imaging 1999;9:468–74.

[8] Schmitt R, Christopoulos G, Meier R, et al. Direct MR arthrography of the wrist in comparison with arthroscopy: a prospective study on 125 patients. Rofo Fortschr Geb Rontgenstr Neuen Bildgeb Verfahr 2003;175:911–9.

[9] Schulte-Altedorneburg G, Gebhard M, Wohlgemuth WA, et al. MR arthrography: pharmacology, efficacy and safety in clinical trials. Skeletal Radiol 2003;32: 1–12.

[10] Schweitzer ME, Brahme SK, Hodler J, et al. Chronic wrist pain: spin-echo and short tau inversion recovery MR imaging and conventional and MR arthrography. Radiology 1992;182:205–11.

[11] Zanetti M, Bram J, Hodler J. Triangular fibrocartilage and intercarpal ligaments of the wrist: does MR

arthrography improve standard MRI? J Magn Reson Imaging 1997;7:590–4.

[12] Grainger AJ, Elliott JM, Campbell RS, et al. Direct MR arthrography: a review of current use. Clin Radiol 2000;55:163–76.

[13] Peh WC, Cassar-Pullicino VN. Magnetic resonance arthrography: current status. Clin Radiol 1999;54: 575–87.

[14] Steinbach LS, Palmer WE, Schweitzer ME. Special focus session: MR arthrography. Radiographics 2002; 22:1223–46.

[15] Linkous MD, Gilula LA. Wrist arthrography today. Radiol Clin North Am 1998;36:651–72.

[16] Brown RR, Clarke DW, Daffner RH. Is a mixture of gadolinium and iodinated contrast material safe during MR arthrography? AJR Am J Roentgenol 2000;175:1087–90.

[17] Theumann NH, Pfirrmann CW, Antonio GE, et al. Extrinsic carpal ligaments: normal MR arthrographic appearance in cadavers. Radiology 2003;226:171–9.

[18] Beaulieu CF, Ladd AL. MR arthrography of the wrist: scanning-room injection of the radiocarpal joint based on clinical landmarks. AJR Am J Roentgenol 1998; 170:606–8.

[19] Yoshioka H, Ueno T, Tanaka T, et al. High-resolution MR imaging of triangular fibrocartilage complex (TFCC): comparison of microscopy coils and a con-ventional small surface coil. Skeletal Radiol 2003; 32:575–81.

[20] Zlatkin MB, Rosner J. MR imaging of ligaments and triangular fibrocartilage complex of the wrist. Magn Reson Imaging Clin N Am 2004;12:301–31.

[21] Yu JS, Habib PA. Normal MR imaging anatomy of the wrist and hand. Magn Reson Imaging Clin N Am 2004;12:207–19.

[22] Stoller D, Brody G. The wrist and hand. In: Stoller D, editor. Magnetic resonance imaging in orthopaedics and sports medicine. Philadelphia: Lippincott-Raven Publishers; 1997. p. 851–993.

[23] Smith DK. Scapholunate interosseous ligament of the wrist: MR appearances in asymptomatic volunteers and arthrographically normal wrists. Radiology 1994; 192:217–21.

[24] Smith DK, Snearly WN. Lunotriquetral interosseous ligament of the wrist: MR appearances in asymptom-atic volunteers and arthrographically normal wrists. Radiology 1994;191:199–202.

[25] Smith DK. MR imaging of normal and injured wrist ligaments. Magn Reson Imaging Clin N Am 1995;3: 229–48.

[26] Totterman SM, Miller R, Wasserman B, et al. Intrinsic and extrinsic carpal ligaments: evaluation by three-dimensional Fourier transform MR imaging. AJR Am J Roentgenol 1993;160:117–23.

[27] Totterman SM, Miller RJ. MR imaging of the triangular fibrocartilage complex. Magn Reson Imag-ing Clin N Am 1995;3:213–28.

[28] Totterman SM, Miller RJ, McCance SE, et al. Lesions of the triangular fibrocartilage complex: MR findings

with a three-dimensional gradient-recalled-echo sequence. Radiology 1996;199:227–32.

[29] Palmer AK, Werner FW. The triangular fibrocartilage complex of the wrist: anatomy and function. J Hand Surg 1981;6:153–62.

[30] Garcia-Elias M. Soft-tissue anatomy and relationships about the distal ulna. Hand Clin 1998;14:165–76.

[31] Thiru RG, Ferlic DC, Clayton ML, et al. Arterial anatomy of the triangular fibrocartilage of the wrist and its surgical significance. J Hand Surg [Am] 1986; 11:258–63.

[32] Benjamin M, Evans EJ, Pemberton DJ. Histological studies on the triangular fibrocartilage complex of the wrist. J Anat 1990;172:59–67.

[33] Chidgey LK, Dell PC, Bittar ES, et al. Histologic anatomy of the triangular fibrocartilage. J Hand Surg [Am] 1991;16:1084–100.

[34] Mikic ZD. Age changes in the triangular fibrocartilage of the wrist joint. J Anat 1978;126:367–84.

[35] Cooney WP, Linscheid RL, Dobyns JH. Triangular fibrocartilage tears. J Hand Surg [Am] 1994;19: 143–54.

[36] Dailey SW, Palmer AK. The role of arthroscopy in the evaluation and treatment of triangular fibrocartilage complex injuries in athletes. Hand Clin 2000;16: 461–76.

[37] Palmer AK. Triangular fibrocartilage complex lesions: a classification. J Hand Surg 1989;14:594–606.

[38] Friedman SL, Palmer AK. The ulnar impaction syndrome. Hand Clin 1991;7:295–310.

[39] Gilula LA, Palmer AK. Is it possible to call a "tear" on arthrograms or magnetic resonance imaging scans? J Hand Surg [Am] 1993;18:547.

[40] Zanetti M, Linkous MD, Gilula LA, et al. Characteristics of triangular fibrocartilage defects in symptomatic and contralateral asymptomatic wrists. Radiology 2000;216:840–5.

[41] Haims AH, Schweitzer ME, Morrison WB, et al. Limitations of MR imaging in the diagnosis of peripheral tears of the triangular fibrocartilage of the wrist. AJR Am J Roentgenol 2002;178:419–22.

[42] Morley J, Bidwell J, Bransby-Zachary M. A comparison of the findings of wrist arthroscopy and magnetic resonance imaging in the investigation of wrist pain. J Hand Surg 2001;26:544–6.

[43] Oneson SR, Scales LM, Erickson SJ, et al. MR imaging of the painful wrist. Radiographics 1996;16: 997–1008.

[44] Oneson SR, Timins ME, Scales LM, et al. MR imaging diagnosis of triangular fibrocartilage pathology with arthroscopic correlation. AJR Am J Roentgenol 1997;168:1513–8.

[45] Potter HG, Asnis-Ernberg L, Weiland AJ, et al. The utility of high-resolution magnetic resonance imaging in the evaluation of the triangular fibrocartilage complex of the wrist. J Bone Joint Surg Am 1997;79: 1675–84.

[46] Bernstein MA, Nagle DJ, Martinez A, et al. A comparison of combined arthroscopic triangular

fibrocartilage complex debridement and arthroscopic wafer distal ulna resection versus arthroscopic triangular fibrocartilage complex debridement and ulnar shortening osteotomy for ulnocarpal abutment syndrome. Arthroscopy 2004;20:392–401.

[47] Geissler WB, Freeland AE, Weiss AP, et al. Techniques of wrist arthroscopy. Instr Course Lect 2000; 49:225–37.

[48] Feldon P, Terrono AL, Belsky MR. The "wafer" procedure: partial distal ulnar resection. Clin Orthop 1992;275:124–9.

[49] Loftus JB. Arthroscopic wafer for ulnar impaction syndrome. Tech Hand Upper Extrem Surg 2000;4: 182–8.

[50] Hermansdorfer JD, Kleinman WB. Management of chronic peripheral tears of the triangular fibrocartilage complex. J Hand Surg 1991;16:340–6.

[51] Minami A, Kato H. Ulnar shortening for triangular fibrocartilage complex tears associated with ulnar positive variance. J Hand Surg [Am] 1998;23:904–8.

[52] Trumble TE, Gilbert M, Vedder N. Ulnar shortening combined with arthroscopic repairs in the delayed management of triangular fibrocartilage complex tears. J Hand Surg [Am] 1997;22:807–13.

[53] Melone Jr CP, Nathan R. Traumatic disruption of the triangular fibrocartilage complex. Pathoanatomy. Clin Orthop 1992;275:65–73.

[54] Scheker LR, Belliappa PP, Acosta R, et al. Reconstruction of the dorsal ligament of the triangular fibrocartilage complex. J Hand Surg 1994;19:310–8.

[55] Jantea CL, Baltzer A, Ruther W. Arthroscopic repair of radial-sided lesions of the triangular fibrocartilage complex. Hand Clin 1995;11:31–6.

[56] Taleisnik J. The ligaments of the wrist. J Hand Surg [Am] 1976;1:110–8.

[57] Mayfield JK. Wrist ligamentous anatomy and pathogenesis of carpal instability. Orthop Clin North Am 1984;15:209–16.

[58] Berger RA. The anatomy of the ligaments of the wrist and distal radioulnar joints. Clin Orthop 2001;383: 32–40.

[59] Cooney WP, Dobyns JH, Linscheid RL. Arthroscopy of the wrist: anatomy and classification of carpal instability. Arthroscopy 1990;6:133–40.

[60] Timins ME, Jahnke JP, Krah SF, et al. MR imaging of the major carpal stabilizing ligaments: normal anatomy and clinical examples. Radiographics 1995;15: 575–87.

[61] Mitsuyasu H, Patterson RM, Shah MA, et al. The role of the dorsal intercarpal ligament in dynamic and static scapholunate instability. J Hand Surg [Am] 2004;29:279–88.

[62] Mizuseki T, Ikuta Y. The dorsal carpal ligaments: their anatomy and function. J Hand Surg 1989;14:91–8.

[63] Viegas SF, Yamaguchi S, Boyd NL, et al. The dorsal ligaments of the wrist: anatomy, mechanical properties, and function. J Hand Surg [Am] 1999;24:456–68.

[64] Berger RA, Kauer JM, Landsmeer JM. Radioscapholunate ligament: a gross anatomic and histologic

study of fetal and adult wrists. J Hand Surg [Am] 1991;16:350–5.

[65] Linscheid RL, Dobyns JH, Beabout JW, et al. Traumatic instability of the wrist: diagnosis, classification, and pathomechanics. J Bone Joint Surg Am 1972;54:1612–32.

[66] Gelberman RH, Cooney III WP, Szabo RM. Carpal instability. Instr Course Lect 2001;50:123–34.

[67] Wright TW, Dobyns JH, Linscheid RL, et al. Carpal instability non-dissociative. J Hand Surg 1994;19:763–73.

[68] Shin SS, Moore DC, McGovern RD, et al. Scapholunate ligament reconstruction using a bone-retinaculum-bone autograft: a biomechanic and histologic study. J Hand Surg [Am] 1998;23:216–21.

[69] Linkous MD, Pierce SD, Gilula LA. Scapholunate ligamentous communicating defects in symptomatic and asymptomatic wrists: characteristics. Radiology 2000;216:846–50.

[70] Moritomo H, Viegas SF, Elder K, et al. The scaphotrapezio-trapezoidal joint. Part 2: A kinematic study. J Hand Surg [Am] 2000;25:911–20.

[71] Watson HK, Ballet FL. The SLAC wrist: scapholunate advanced collapse pattern of degenerative arthritis. J Hand Surg [Am] 1984;9:358–65.

[72] Adolfsson L. Arthroscopic diagnosis of ligament lesions of the wrist. J Hand Surg 1994;19:505–12.

[73] Blatt G. Capsulodesis in reconstructive hand surgery: dorsal capsulodesis for the unstable scaphoid and volar capsulodesis following excision of the distal ulna. Hand Clin 1987;3:81–102.

[74] Walsh JJ, Berger RA, Cooney WP. Current status of scapholunate interosseous ligament injuries. J Am Acad Orthop Surg 2002;10:32–42.

[75] Shin A, Deitch M, Sachar K, et al. Ulnar-sided wrist pain: diagnosis and treatment. J Bone Joint Surg Am 2004;86A:1560–74.

[76] Weiss LE, Taras JS, Sweet S, et al. Lunotriquetral injuries in the athlete. Hand Clin 2000;16:433–8.

[77] Shin AY, Weinstein LP, Berger RA, et al. Treatment of isolated injuries of the lunotriquetral ligament: a comparison of arthrodesis, ligament reconstruction and ligament repair. J Bone Joint Surg Br 2001;83:1023–8.

[78] Brown DE, Lichtman DM. Midcarpal instability. Hand Clin 1987;3:135–40.

[79] Cerezal L, del Pinal F, Abascal F, et al. Imaging findings in ulnar-sided wrist impaction syndromes. Radiographics 2002;22:105–21.

[80] Cerezal L, del Pinal F, Abascal F. MR imaging findings in ulnar-sided wrist impaction syndromes. Magn Reson Imaging Clin N Am 2004;12:281–99.

[81] Tomaino MM. Ulnar impaction syndrome in the ulnar negative and neutral wrist: diagnosis and pathoanatomy. J Hand Surg 1998;23:754–7.

[82] Imaeda T, Nakamura R, Shionoya K, et al. Ulnar impaction syndrome: MR imaging findings. Radiology 1996;201:495–500.

[83] Tomaino MM, Weiser RW. Combined arthroscopic TFCC debridement and wafer resection of the distal ulna in wrists with triangular fibrocartilage complex tears and positive ulnar variance. J Hand Surg [Am] 2001;26:1047–52.

[84] Tomaino MM, Gainer M, Towers JD. Carpal impaction with the ulnar styloid process: treatment with partial styloid resection. J Hand Surg 2001;26:252–5.

[85] Topper SM, Wood MB, Ruby LK. Ulnar styloid impaction syndrome. J Hand Surg 1997;22:699–704.

[86] Garcia-Elias M. Dorsal fractures of the triquetrum-avulsion or compression fractures? J Hand Surg 1987;12:266–8.

[87] Hauck RM, Skahen III J, Palmer AK. Classification and treatment of ulnar styloid nonunion. J Hand Surg 1996;21:418–22.

[88] Malik AM, Schweitzer ME, Culp RW, et al. MR imaging of the type II lunate bone: frequency, extent, and associated findings. AJR Am J Roentgenol 1999;173:335–8.

[89] Pfirrmann CW, Theumann NH, Chung CB, et al. The hamatolunate facet: characterization and association with cartilage lesions: magnetic resonance arthrography and anatomic correlation in cadaveric wrists. Skeletal Radiol 2002;31:451–6.

[90] Thurston AJ, Stanley JK. Hamato-lunate impingement: an uncommon cause of ulnar-sided wrist pain. Arthroscopy 2000;16:540–4.

[91] Harper MT, Chandnani VP, Spaeth J, et al. Gamekeeper thumb: diagnosis of ulnar collateral ligament injury using magnetic resonance imaging, magnetic resonance arthrography and stress radiography. J Magn Reson Imaging 1996;6:322–8.

[92] Plancher KD, Ho CP, Cofield SS, et al. Role of MR imaging in the management of "skier's thumb" injuries. Magn Reson Imaging Clin N Am 1999;7:73–84.

[93] Melone Jr CP, Beldner S, Basuk RS. Thumb collateral ligament injuries: an anatomic basis for treatment. Hand Clin 2000;16:345–57.

[94] Fairhurst M, Hansen L. Treatment of "gamekeeper's thumb" by reconstruction of the ulnar collateral ligament. J Hand Surg 2002;27:542–5.

[95] Barron OA, Glickel SZ, Eaton RG. Basal joint arthritis of the thumb. J Am Acad Orthop Surg 2000;8:314–23.

[96] Masson JA, Golimbu CN, Grossman JA. MR imaging of the metacarpophalangeal joints. Magn Reson Imaging Clin N Am 1995;3:313–25.

[97] Pomerance JF. Painful basal joint arthritis of the thumb. Part II: Treatment. Am J Orthop 1995;24:466–72.

[98] Bergin D, Schweitzer ME. Indirect magnetic resonance arthrography. Skeletal Radiol 2003;32:551–8.

[99] Haims AH, Schweitzer ME, Morrison WB, et al. Internal derangement of the wrist: indirect MR arthrography versus unenhanced MR imaging. Radiology 2003;227:701–7.

[100] Vahlensieck M, Peterfy CG, Wischer T, et al. Indirect MR arthrography: optimization and clinical applications. Radiology 1996;200:249–54.

ELSEVIER
SAUNDERS

Radiol Clin N Am 43 (2005) 733 – 746

RADIOLOGIC
CLINICS
of North America

MR Arthrography of the Knee: How, Why, When

Christine B. Chung, MD[a,*], Ilma L. Isaza, MD[b], Maritza Angulo, MD[c],
Ron Boucher, MD[d], Tudor Hughes, MD[a]

[a]Department of Radiology, University of California San Diego and Veterans Affairs Healthcare System,
3350 La Jolla Village Drive, La Jolla, CA 92161, USA
[b]CT Scanner de Mexico, Puebla No. 228, Colonia Roma, CP 06700, Mexico
[c]Fundacion Clinica Medica Sur, Calle Puente de Piedra No. 150, Delegacion Tlalpan, Mexico
[d]Naval Medical Center, 34800 Bob Wilson Drive, Suite 204, San Diego, CA 92134–1204, USA

Direct MR arthrography is an imaging examination that combines the injection of a saline or dilute gadolinium solution into an articulation, followed by MR imaging of that articulation. By distending the joint, intra-articular structures are optimally visualized and capsular integrity can be assessed by noting the distribution of fluid within the joint. Although this imaging technique has been increasingly used to evaluate pathology in several articulations, it is the glenohumeral joint and knee that have been most frequently interrogated by MR arthrography.

To date, the primary indications for MR arthrography of the knee are twofold: the first includes the evaluation of the postoperative meniscus; the second involves the diagnosis and characterization of osteochondral lesions. Although arthrography introduces an invasive component to the imaging evaluation of the knee, it is an established method for optimizing diagnostic accuracy. Arthrography of the knee is well-tolerated by most patients and has proved to be useful when considering treatment planning [1]. It is most accurate and precise when internal derangement is suspected based on clinical history, symptomatology, and physical examination findings. This article focuses on the technical approach for MR arthrography of the knee including its potential compli-

cations and emphasizes the pathology whose visualization is optimized by this imaging technique. The latter includes established indications in the imaging literature, and anecdotal situations in which MR arthrography has proved useful in the setting of diagnostic dilemmas.

Technique

Contrast agents for MR arthrography include a dilute solution of paramagnetic contrast material, normal saline, or Ringer's solution [2]. Binkert et al [2] found comparable diagnostic accuracy among differing contrast agents, but preferred the image quality unique to the gadopentetate dimeglumine–enhanced arthrogram. Gadolinium is first diluted with 1 mL of gadolinium in 200 mL of normal saline (0.9%). The suggested concentration of a gadopentetate dimeglumine solution is 2 mmol/L [3,4]. Iodinated contrast can be included in the solution, and is advocated in some cases because it allows confirmation of initial needle placement, verifies an entirely intra-articular injection, and allows assessment of the distribution of fluid within the articulation. The 2 mmol/L concentration in this case is achieved by preparing a solution of 2 mL of gadolinium in 200 mL of normal saline. Equal parts of this solution and iodinated contrast material result in a 1:200 dilution ratio.

* Corresponding author.
  E-mail address: cbchung@ucsd.edu (C.B. Chung).

Although classically administered under fluoroscopic guidance, intra-articular gadolinium can also be instilled under sonographic or MR imaging control and reports of the benefits of each imaging method can be found in the literature [5–7]. Because of the superficial nature of this articulation, the injection can also be accomplished without imaging guidance [8].

The patient is placed in supine position with a small pillow or towel placed under the knee. This results in slight flexion of the knee, promoting deactivation of the extensor mechanism. Classically, a standard 20-gauge needle is placed within the medial or lateral patellofemoral joint while downward pressure is placed on the side of the patella opposite needle entry. Balloting the patella in this fashion allows opening of the opposite joint space and facilitates unhindered needle entry. A second technique for intra-articular injection of the knee uses an infrapatellar needle entry site on either side of the patellar tendon (Fig. 1). If this approach is used, care must be taken to insert the needle far enough into the joint to extend beyond the inner margin of Hoffa's fat pad. The intra-articular position of the needle tip is confirmed either by a test injection of iodinated contrast agent, or injection of a solution that includes iodinated contrast material as previously mentioned. Alternatively, intra-articular position can be confirmed by the aspiration of joint fluid. In either case, any excess joint effusion within the articulation should be aspirated before the injection of the dilute

gadolinium solution to maintain a standard concentration of contrast within the joint [9]. In the assessment of the initial injection, the contrast agent should flow away from the needle tip and outline the articular surface of the femorotibial and patellofemoral joints. Careful injection of the diluted contrast solution is performed until slight resistance is felt. Although overdistention of the joint may lead to extravasation of contrast material in other articulations, such as the glenohumeral joint, resulting in incomplete articular distention and suboptimal visualization of intra-articular structures, the capacity of the knee joint is quite large and extravasation from iatrogenic capsular rupture is a rare occurrence. In the authors' experience the optimal total injected volume is approximately 35 to 50 mL. After the dilute contrast solution has been placed in the joint, a tourniquet can be secured around the superior aspect of the joint to restrict preferential filling of the suprapatellar pouch [8]. MR imaging should not be delayed for more than 1 hour so as to avoid absorption of the gadolinium by the synovium [10]. If there is no expectation of an extended time interval between arthrogram and MR imaging, then there is limited utility in administering epinephrine, which is noted to prevent synovial absorption of gadolinium.

Axial, oblique coronal, and oblique sagittal plane images are acquired with a T1-weighted fat-suppressed sequence using a dedicated extremity coil. Fat suppression is an important tool because it eliminates the signal from periarticular fat. Diagnostic accuracy is increased with the use of fat suppression [11,12]. Partial- and full-thickness cuff tears may not be distinguishable on standard T1-weighted images because of the similar signal intensities between fat and gadolinium in the absence of fat-suppression. With T1-weighted fat-suppressed images, the signal intensity of the contrast remains unchanged, whereas the fat is low in signal intensity, allowing facile distinction between the two. Oblique coronal fast spin echo T2-weighted fat-suppressed or fast multiplanar inversion recovery sequences are also performed in conjunction with the T1-weighted images (Fig. 2). This allows detection of areas of edema that would otherwise be overlooked with T1-weighted images alone [13].

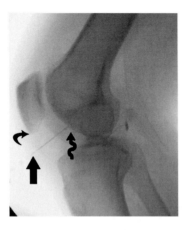

Fig. 1. Lateral projection image obtained during fluoroscopic guidance for an anterior approach infrapatellar knee injection shows the needle entry (*straight arrow*) below the inferior pole of the patella (*curved arrow*). Note that the needle extends well into the joint (*wavy arrow*) to avoid injection into Hoffa's fat pad.

Technical complications

With the proper knowledge of both arthrographic and imaging pitfalls, most of the potential limitations can be anticipated and easily overcome.

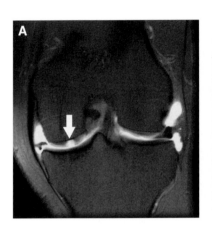
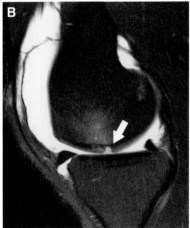

Fig. 2. (*A*) Coronal T1-weighted fat-suppressed MR arthrogram image shows contrast outlining a chondral lesion on the weight-bearing surface of the medial femoral condyle (*arrow*). The vertical margins of the lesion suggest it is acute in nature. (*B*) Sagittal T2-weighted fat-suppressed MR arthrogram image profiles the lesion and adjacent bone marrow edema, supporting the theory and clinical history of an acute injury in this patient who had undergone previous meniscectomy.

Gadolinium concentration

Gadolinium must be diluted, usually with normal saline, to a concentration between 2 and 4 mmol/L to achieve diagnostic imaging, as previously mentioned. If gadolinium is not properly diluted, then a magnetic susceptibility artifact results. The imaging appearance is that of diffuse low signal intensity in the distribution of the injected contrast material (Fig. 3). This poses no danger to the patient; however, it invalidates the imaging examination because all intra-articular structures are effaced by the diffuse low signal intensity [13].

Contrast extravasation

Contrast extravasation can occur either secondary to overdistention of the knee joint with contrast and subsequent capsular rupture or, more commonly, by extra-articular needle placement with injection of contrast. Exercise, which was one of the hallmarks of CT arthrography, has not been recommended for

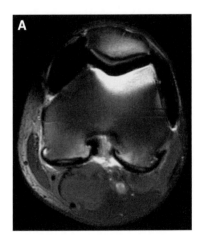
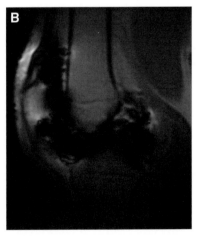

Fig. 3. Axial (*A*) and sagittal (*B*) T1-weighted fat-suppressed MR arthrogram images after unintentional concentrated gadolinium solution was injected into the knee show diffuse low signal intensity throughout the joint. The low signal intensity precludes visualization of all intra-articular structures.

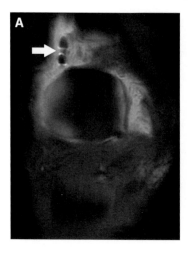

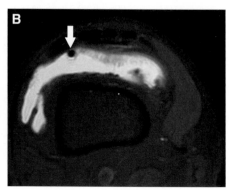

Fig. 4. Coronal (*A*) and axial (*B*) T1-weighted fat-suppressed MR arthrogram images show air bubbles at the anterior aspect of the joint mimicking intra-articular bodies (*arrow*).

MR arthrography in joints such as the shoulder, because it hastens the capsular absorption of gadolinium and may result in extravasation [14]. In the knee, however, where constant pressure from weight bearing may compress abnormalities in the menisci, gentle exercise in the form of walking may help to increase intra-articular pressure driving contrast into meniscal lesions.

Introduction of intra-articular air

Unintentional introduction of air into the articulation with injection may simulate intra-articular osseous bodies (Fig. 4). These air bubbles can occasionally be distinguished from osseous bodies because of the tendency of air preferentially to localize in the nondependent portion of the articulation. If the MR arthrogram study is being monitored and the presence of air bubbles is detected and simulating pathology, the patient can be placed in prone position and reimaged to mobilize the air within the articulation.

Fat suppression

The use of fat suppression in conjunction with T1-weighted imaging for MR arthrography increases both the sensitivity and specificity of this imaging technique. Heterogeneous fat suppression can impair the detection of labral tears because of difficulty in the distinction between peribursal fat and contrast agent. In addition, it can simulate bone marrow signal changes, such as contusion [15].

Contrast resorption

A delay in the time interval between gadolinium arthrography and subsequent MR imaging can result in significant synovial resorption of contrast agent from the articulation (Fig. 5). The highly vascular synovium has a role in restoring the joint fluid equilibrium and attempts to maintain the intra-articular physiologic volume and concentration. An extended interval between arthrography and imaging allows this process to occur and limits the interpre-

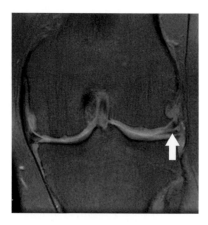

Fig. 5. Coronal T1-weighted fat-suppressed MR arthrogram image in a patient whose MR imaging study was delayed over 2 hours after injection of a dilute gadolinium solution. Much of the contrast has been resorbed from the articulation. There is, however, the suggestion of gadolinium collecting in the postoperative lateral meniscus (*arrow*), suggesting the presence of a recurrent meniscal tear.

tative functionality of the study. Wagner et al [10] reports an acceptable delay of up to 1 hour for MR arthrography of the shoulder. In the experience of the authors, a similar time frame can be applied to other articulations.

Applications

Postoperative knee

The postoperative knee presents with its own set of clinical and imaging diagnostic challenges. Although one of the basic and universally accepted advantages of MR imaging is excellent soft tissue contrast, in the postoperative knee the presence of redundant capsular tissue, synovial hypertrophy, variations in signal intensity of tissue in various stages of healing and injury, and a paucity of intra-articular fluid in many cases make the accurate diagnosis and characterization of pathology difficult. The presence of intra-articular contrast material not only provides physical separation of adjacent apposed structures allowing delineation of one structure from the next, but also allows further characterization of signal intensity abnormalities within structures.

Postoperative meniscus

Standard MR imaging has proved extremely valuable in the preoperative diagnosis of meniscal tears, but has been reported to have unreliable results in the postoperative meniscus with accuracies ranging between 38% and 82% [16–18]. The accurate detection of recurrent meniscal tear in the postoperative knee has become of paramount importance because the standard of treatment for symptomatic meniscal tears has shifted from partial or complete meniscectomy to meniscal-preserving procedures including primary repair with suture material or bioabsorbable arrows [16,19–21]. The conventional diagnostic criteria used to diagnose meniscal tears are difficult to apply to the postoperative meniscus for several reasons. When a partial meniscectomy is performed, hyperintense intrasubstance signal intensity can be converted into signal intensity that extends to an articular surface, simulating a meniscal tear. In addition, in the setting of a primary meniscal repair, a healing tear may be seen as hyperintense signal intensity in the meniscal remnant. The high signal component in this case is caused by the presence of granulation tissue [16]. Other considerations that affect the ability to detect abnormalities in the postoperative meniscus include the size of the original

tear and, in the setting of meniscectomy, the extent of resection. Clearly, the more meniscal tissue that remains, the easier it is to diagnose pathology in the remnant.

The diagnostic criteria for recurrent meniscal tear in the postoperative setting differ from the virgin meniscus and do so with both standard MR imaging and MR arthrography. The classic grade III meniscal signal intensity, linear signal intensity extending to an articular surface, is known to be present in the stable, healed, postoperative meniscus [22–25]. It is generally agreed that meniscal instability and recurrent tear in the postoperative meniscus are diagnosed by the presence of a meniscal fragment separate from the residual meniscus on standard or MR arthrogram images, intrameniscal signal intensity equal in brightness to that of joint fluid on standard T2-weighted MR images extending to an articular surface, or bright intrameniscal signal intensity extending to an articular surface on T1-weighted fat-suppressed MR arthrogram images of the knee (Fig. 6) [16,17,22, 23,25,26]. Some authors include irregular meniscal contour as one of the diagnostic criteria for recurrent tear in the postoperative meniscus [16]. In general, this imaging finding, on both standard MR imaging and MR arthrography, is fraught with uncertainty. Smith and Totty [22] found that 30% of patients with extensive meniscal resection presented with intrameniscal signal heterogeneity and marked surface irregularity in the meniscal remnant, most of which were normal by arthroscopic correlation. They concluded that the diagnosis of recurrent tears in the postoperative meniscus with marked contour irregularity must be made with caution because this may be a normal postoperative appearance after meniscectomy. This theory was further corroborated by White et al [26], who reported morphologic changes of the postoperative meniscus to be the least accurate imaging finding for the diagnosis of meniscal retear for both standard MR imaging and MR arthrography.

Although it is generally accepted that MR arthrography is more accurate than MR imaging for the detection of recurrent tear in the postoperative meniscus, White et al [26] found only a small, statistically insignificant increase in accuracy with MR arthrography over standard MR imaging. Applegate et al [23] compared standard MR imaging with MR arthrography in the evaluation of the postoperative meniscus in 37 patients. In menisci that had less than 25% of its substance resected, no statistically significant difference between the diagnostic accuracy of the two imaging methods was noted (accuracy = 89%). In menisci with greater than 25% resection, however, the accuracy of MR

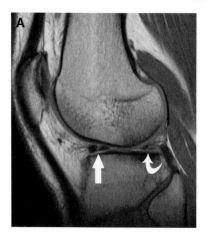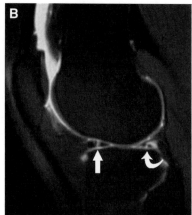

Fig. 6. (*A*) Sagittal proton density weighted standard MR image shows linear signal intensity extending to the articular surfaces of the anterior (*straight arrow*) and posterior (*curved arrow*) horns of the postoperative lateral meniscus. (*B*) Sagittal T1-weighted fat-suppressed MR arthrogram image shows corresponding contrast material entering both the anterior (*straight arrow*) and posterior (*curved arrow*) horns of the lateral meniscus consistent with recurrent meniscal tears.

arthrography (85%) was statistically significantly higher than that of standard MR imaging (65%). Similar results were reported by Magee et al [27], who assessed 100 consecutive MR arthrograms in patients with previous knee surgeries to identify the patient population that benefits the most from MR arthrography of the knee. They concluded, similar to Applegate et al [23], that all patients with meniscal repair and those with greater than 25% meniscal resection benefit from MR arthrographic assessment.

The increased accuracy of MR arthrography in the detection of recurrent meniscal tear in the postoperative meniscus has been attributed to two factors. First, it has been noted that a large volume of fluid in the joint increases intra-articular hydrostatic pressure resulting in separation of the menisci and adjacent articular cartilage. This physical separation obviates the problems of volume averaging of the two tissues, and allows the contrast placed into the articulation to have access to, and track into, potential defects in the meniscal remnant. Second, it has been emphasized that the superior contrast afforded by the combination of T1-weighted fat-suppressed images and a dilute gadolinium solution allows distinction between meniscal retear and granulation tissue or fibrovascular scar. In the former, the gadolinium solution has very bright signal intensity that tracks into the meniscal remnant, whereas granulation tissue and scar are low in signal intensity on T1-weighted fat-suppressed images. On the contrary, with conventional T2-weighted MR images joint fluid approximates the signal intensity of granulation and scar tissue, making it difficult to distinguish between these

entities. Applegate et al [23] did encounter two false-negative diagnoses of healing menisci with MR arthrography. This was attributed to the presence of thin layers of scar tissue covering the surface communication of the meniscal remnant tear, precluding the intrameniscal flow of contrast solution. To further aid in the diagnostic accuracy of MR arthrography for the detection of subtle recurrent meniscal tears, a prearthrogram T1-weighted fat-suppressed image could be obtained. This allows comparison of prearthrogram and postarthrogram images and an opportunity directly to compare signal changes in the meniscus to ascertain whether even trace amounts of contrast enter the meniscal substance (Fig. 7). Although this may not be possible in many cases because of scheduling issues, it can prove helpful in difficult cases.

Postoperative anterior cruciate ligament reconstruction

Studies describing the MR imaging of anterior cruciate ligament (ACL) reconstruction have been limited to small patient population with arthroscopic correlation [28–30]. These studies have shown variable accuracies for standard MR imaging, with the largest study of 16 patients showing 50% sensitivity and 100% specificity for the detection of ACL graft tear [28]. Moreover, it has been established that the distinction between ACL rupture and graft impingement can be quite difficult, if not impossible, to make [31,32]. McCauley et al [29] retrospectively identified 27 patients with ACL reconstruction who had undergone MR arthrography followed by arthroscopy

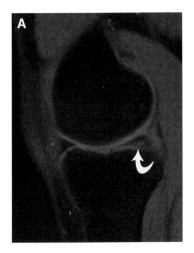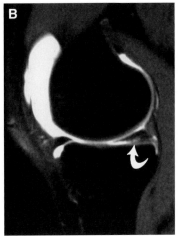

Fig. 7. Sagittal T1-weighted fat-suppressed standard MR image (*A*) can be compared with the corresponding T1-weighted fat-suppressed MR arthrogram image (*B*) if there is uncertainty as to the degree of increased signal (*arrow*) on MR arthrography studies. If the postarthrogram signal is higher than that of the standard MR imaging study, it signals the presence of contrast material in the area in question, in this case representing a recurrent tear in a postoperative meniscus.

within 1 year of the ACL repair. Graft tears were identified with 100% sensitivity by all reviewers, and a specificity ranging from 89% to 100%. Interobserver agreement among three readers was excellent for detection of graft tear (kappa = 0.83–0.92).

The diagnosis of a torn graft was made when the graft fibers could not be identified extending from the femoral tunnel to the tibial tunnel, especially when contrast material extended through a discontinuity in the graft fibers (Fig. 8). Anterior tibial translation and an uncovered posterior horn of the lateral meniscus were specific although insensitive findings of an ACL

graft rupture. The study did not compare standard MR imaging with MR arthrography, but did conclude that MR arthrography can accurately depict the presence of ACL graft tears [29]. In the experience of the authors, a torn ACL graft in the subacute or chronic setting can be quite difficult to detect. In the setting of mid-substance failure, the torn ends of the graft can reside adjacent to each other making discontinuity difficult to detect. In addition, in some cases partial resynovialization around the cruciate ligaments can mimic an intact graft, as can the ligamentum mucosum (also known as the *inferior plica*)

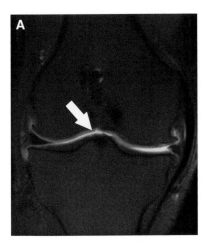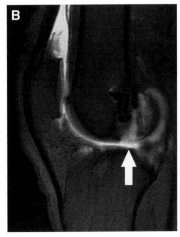

Fig. 8. Coronal (*A*) and sagittal (*B*) T1-weighted fat-suppressed MR arthrogram images in a patient with an ACL reconstruction show discontinuity of the ACL fibers (*arrow*) consistent with a rupture of the graft.

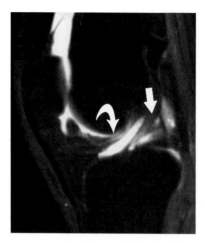

Fig. 9. Sagittal T1-weighted fat-suppressed MR arthrogram image demonstrates the advantage of contrast in the articulation to allow separation of the tissue of the ligamentum mucosum (inferior plica) (*curved arrow*) from that of the ACL (*straight arrow*).

(Fig. 9). Fluid in the joint can help to identify the course and morphology of the ACL graft and the surrounding synovial tissues (Fig. 10).

Cyclops lesion

Arthrofibrosis is characterized by synovial hyperplasia with excessive production of fibrous tissue and inflammatory cell proliferation around the ACL graft [33]. The *cyclops lesion* is a localized type of anterior arthrofibrosis that is observed in 1% to 10% of ACL reconstructions and is considered the second most common cause of loss of terminal extension of the knee, after graft impingement [32]. At arthroscopy,

the cyclops lesion is seen as a head-like fibrous lesion with reddish blue areas of discoloration resembling an eye [32,34,35]. On MR imaging the cyclops lesion is seen as a well-circumscribed nodule with intermediate to low signal intensity on T1-weighted and intermediate-weighted images. The nodule lies in the intercondylar notch, just anterior to the tibial insertion of the graft and posterior to the infrapatellar fat pad. Arthrofibrosis should be distinguished from the fibrotic scars, which are related to previous arthroscopy and are usually seen in the infrapatellar fat pad at the entrance portals. Fibrotic scars are usually linear in nature with horizontal hypointense strands, in contrast to the mass-like appearance of diffuse arthrofibrosis. On some occasions, however, the innermost tip of Hoffa's fat pad can become diffusely low in signal intensity because of scarring from previous surgery. Because of the intimate spatial relationship of Hoffa's fat pad and the ACL, these structures can be difficult to distinguish from one another. In the absence of fluid within the articulation, changes in Hoffa's fat pad can mask the presence of a cyclops lesion. Hoffa's disease can also be observed after ACL reconstruction, presumably related to an inflammatory response to debris or injury of the infrapatellar fat pad [32]. MR imaging demonstrates a hypertrophic and edematous fat body with increased signal on fat-suppressed T2 or STIR images that equals the signal intensity of fluid. The high signal intensity of the fat pad in Hoffa's disease should allow distinction from the fibrous tissue of a cyclops lesion, but it is evident that contrast or fluid in the joint can also be helpful in the physical separation of the fat pad and adjacent fibrosis (Fig. 11). McCauley et al [29] found that, even in the setting of

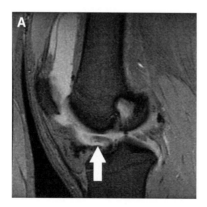
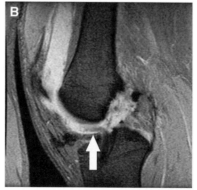

Fig. 10. (*A, B*) Sequential sagittal T1-weighted fat-suppressed MR arthrogram images in a patient who had undergone an ACL reconstruction show a rupture of the femoral attachment of the graft with anterior displacement of the graft fibers (*arrow*) consistent with free stump impingement. The presence of contrast within the joint helps to distinguish the displaced ACL fibers from Hoffa's fat pad.

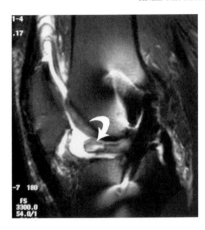

Fig. 11. Sagittal T1-weighted fat-suppressed MR arthrogram image in a patient who had undergone previous ACL reconstruction, presenting with decreased range of motion and a locking sensation, shows contrast separating Hoffa's fat pad from a focal area of synovial proliferation (*arrow*) anterior to the ACL. This is a classic location and imaging appearance for focal arthrofibrosis or the Cyclops lesion.

MR arthrography, both localized anterior arthrofibrosis and graft impingement were difficult to detect and showed significant interobserver variability in diagnosis. This is likely a reflection of the combination of lack of familiarity and difficulty in diagnosis of these entities.

Osteochondral lesions

An osteochondral lesion involves the separation of a segment of articular cartilage along with its underlying bone. It is a general term used to characterize pathology encountered at the articular surface of a joint with no allusion to etiology. Mechanically, an osteochondral lesion is produced by either repetitive and prolonged joint overloading or a sudden impact that results in high compressive stress to the tissue and high shear stress at the subchondral bone junction. This mechanical stress can occur in the setting of normal or abnormal underlying bone. Differential considerations regarding etiology include the posttraumatic knee, osteochondritis dissecans, and the insufficiency fracture of spontaneous osteonecrosis of the knee. The distinction between these entities is made through a combination of clinical and imaging findings including patient demographics, presenting history, location of lesion, and associated structural abnormalities.

The goal of imaging is first to diagnose the abnormality; then characterize it with regard to degree of involvement, stability of the lesion, and complica-tions associated with the lesion; and finally to offer imaging information that may suggest acuity of the injury and etiology. It is MR imaging that has emerged as the most sensitive and reliable means for the assessment of osteochondral lesions [36–40]. The MR imaging diagnosis is made by identifying signal change or morphology abnormality at the articular surface. In general, signal intensity changes in the subchondral bone that are low on T1-weighted images without fragmentation or high signal intensity at the interface of the lesion to normal bone on fluid-sensitive sequences are considered to be stable in nature. De Smet et al [39] showed that the presence of any one of four MR imaging findings indicates an unstable osteochondral lesion. These findings include a line of high signal deep to the fragment as seen on T2-weighted images, an articular fracture indicated by high signal passing through the subchondral bone plate, a focal osteochondral defect, and a 5-mm diameter fluid-filled cyst deep to the lesion. Other authors have suggested that the presence of a bright line deep to the osteochondral fragment may result from granulation tissue and that a break in the articular surface must be demonstrated with MR imaging to be sure that the fragment is unstable [36,38]. Kramer et al [41] compared standard MR imaging with MR arthrography in 25 knees with suspected osteochondral lesion. On conventional images, abnormalities were correctly detected and staged in 39% of T1-weighted spin echo and 57% of gradient echo images. After intra-articular contrast injection of a dilute gadolinium solution, correct diagnoses improved to 93% on T1-weighted images and 100% on gradient echo images. MR arthrographic images were superior to standard MR images because large amounts of fluid could be introduced into the articulation, and the contrast solution was more visible than effusion in the chondral defect and in the interface between lesion and native bone, the primary findings used to establish an unstable lesion. The use of contrast material and T1-weighting also obviates the confusion of distinguishing between fluid and granulation tissue on fluid-sensitive sequences (Fig. 12). A fluid-sensitive sequence should accompany the T1-weighted fat-suppressed MR arthrographic images to allow detection of subchondral cyst formation or bone marrow edema in or around the osteochondral lesion.

Chondral lesions

Several studies have compared conventional with arthrographic MR imaging in the detection and staging of chondral defects [42–45]. Reports have

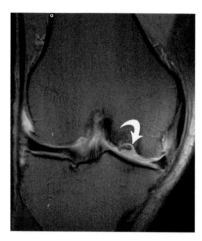

Fig. 12. Coronal T1-weighted fat-suppressed MR arthrogram image in a patient with osteochondritis dissecans shows contrast material (*arrow*) extending between the osteochondral fragment and the native bone consistent with an unstable osteochondral fragment.

shown superior sensitivities and specificities for MR arthrography in assessing intermediate stages of cartilage lesions [46]. Contrast solution outlines the cartilage defect and allows separation of the articular surfaces from each other and from adjacent meniscal tissue, making the lesions more conspicuous. By outlining the defect, more information regarding potential etiology can be ascertained from the image. In general, sharp, vertical margins of a lesion on the weight-bearing surface of an articulation accompanied by adjacent bone marrow edema suggest an acute lesion (see Fig. 2). On the contrary, however, shallow lesions with horizontal, wide margins suggest a more chronic, degenerative process (Fig. 13).

In the past, it was believed that MR arthrography permitted a higher level of confidence in the identification of chondral lesions than CT arthrography [43,45]. With the advent of multidetector CT, however, both imaging methods seem to be equal in accuracy for the detection of intermediate- and high-grade chondral lesions [47,48]. The disadvantage of CT arthrography as compared with MR arthrography remains the lack of soft tissue contrast, and the loss of additional information afforded by the MR imaging technique.

Results of several experimental studies have indicated that T1-weighted MR imaging enhanced by intra-articular gadolinium dimeglumine by means of diffusion has the potential to demonstrate structural abnormalities of hyaline cartilage [7,49–51]. This technique has not yet been evaluated sufficiently in vitro or in vivo to be introduced in the clinical setting; however, it does hold the potential to detect the earliest stages of cartilage degeneration. If successful, this technique could revolutionize the early detection of cartilage degeneration, provide a means to monitor the natural progression of cartilage degeneration, and a means to monitor the efficacy of therapeutic intervention.

Intra-articular bodies

Any process that leads to damage of the articular surface of the knee may result in the formation of cartilaginous or osteocartilaginous bodies. Intra-articular bodies have been categorized based on origin: articular, synovial, or traumatic disease [52]. Cartilaginous or osteocartilaginous bodies may remain in situ, either partially detached or completely detached. When free, they can migrate and often collect in the intercondylar notch, beneath the medial collateral ligament, along the course of the popliteal tendon sheath, or in a Baker's cyst. If left in the joint, bodies can grow because of the repetitive deposition of layers of fibrous and cartilaginous tissue [52].

Brossmann et al [53] explored the role of CT, computed arthrotomography, MR imaging, and MR arthrography for the detection of osseous and cartilaginous intra-articular bodies in the knee. They concluded that MR arthrography was the best imaging technique for detection of individual intra-

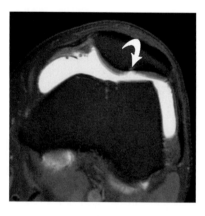

Fig. 13. Axial T1-weighted fat-suppressed MR arthrogram image shows contrast outlining a chondral lesion (*arrow*) of the lateral facet of the patella that seems chronic in nature. The contrast in the patellofemoral joint separates the articular surfaces and outlines the horizontal margins of the chondral lesion. The latter finding suggests that the lesion is chronic in nature.

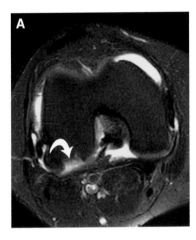
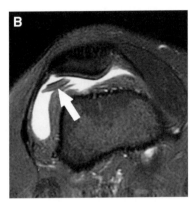

Fig. 14. (*A*) Axial T1-weighted fat-suppressed MR arthrogram image just above the level of the joint line shows the donor site (*curved arrow*) for an osteochondral lesion. (*B*) A more superior axial T1-weighted fat-suppressed MR arthrogram image shows the displaced osteochondral fragment (*straight arrow*) lodged in the lateral patellofemoral joint. The contrast in the articulation allows separation of the articular surfaces and surrounds the intra-articular body making it more evident.

articular bodies, emphasizing the spoiled GRASS (gradient recalled acquisition in steady state) and T2-weighted sequences as the most accurate imaging sequences for this purpose. The accuracy of MR arthrography (92%) in the detection of osseous and cartilaginous bodies proved significantly better than that of standard MR imaging (57%–92%). Not only can MR arthrography delineate loose cartilage and bone in the joint, it has the capability of identifying the donor site for the body and associated soft tissue findings that may exist (Fig. 14).

Cystic adventitial disease

Cystic adventitial disease of the popliteal artery is an unusual condition of uncertain etiology, in which a mucin-containing cyst forms in the wall of the artery and produces lower-extremity claudication, typically in young and middle-aged men. Two plausible theories for this process have been presented in the literature. The adventitial cysts often contain crystal-clear fluid, similar in chemical content to the fluid in ganglion cysts. The presence of mucin-secreting cells

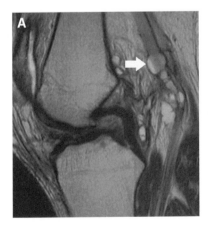
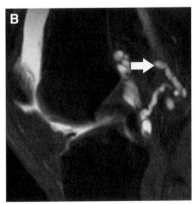

Fig. 15. (*A*) Sagittal proton density weighted standard MR image in a patient who presented with posterior fossa pain and claudication shows multiple cystic structures adjacent to the posterior capsule of the knee joint and adjacent to the posterior fossa vessels (*arrow*). (*B*) Corresponding sagittal T1-weighted fat-suppressed MR arthrogram image demonstrates contrast material emanating from the joint into the cystic structures (*arrow*) noted on the prearthrogram study. These findings are consistent with cystic adventitial disease of the popliteal artery.

in the adventitial wall of the vessel may result either from aberrant developmental inclusion within the adventitia, or the tracking of capsular synovial cysts from a neighboring joint. The latter proposal is supported by surgical evidence of communication between these cystic lesions and the joint capsule of the knee [54,55]. MR imaging may facilitate the diagnosis of cystic adventitial disease by revealing cyst-like structures closely invested in a compressed artery [56]. MR arthrography, although not described in the literature, can establish the diagnosis of this entity by documenting the communication of contrast from the articulation to the surrounding vessel wall (Fig. 15). The diagnosis is important because cyst excision is associated with an excellent prognosis. CT-directed percutaneous aspiration has been performed, but has been associated with recurrences [55,57].

Summary

Direct MR arthrography of the knee is an invaluable study that helps to improve not only the diagnostic sensitivity, specificity, and accuracy for pathologic processes in the knee, but also proves instrumental in providing a detailed characterization of pathology that is crucial for treatment planning. Indications for the study include the evaluation of the postoperative knee; diagnosis and characterization of osteochondral and chondral lesions; diagnosis of intra-articular bodies; and diagnosis of less common entities, such as cystic adventitial disease about this articulation.

References

[1] Binkert CA, Zanetti M, Hodler J. Patient's assessment of discomfort during MR arthrography of the shoulder. Radiology 2001;221:775–8.

[2] Binkert CA, Zanetti M, Gerber C, et al. MR arthrography of the glenohumeral joint: two concentrations of gadoteridol versus Ringer solution as the intraarticular contrast material. Radiology 2001;220: 219–24.

[3] Engel A. Magnetic resonance knee arthrography: enhanced contrast by gadolinium complex in the rabbit and in humans. Acta Orthop Scand Suppl 1990;240: 1–57.

[4] Kopka L, Funke M, Fischer U, et al. MR arthrography of the shoulder with gadopentetate dimeglumine: influence of concentration, iodinated contrast material, and time on signal intensity. AJR Am J Roentgenol 1994;163:621–3.

[5] Petersilge CA, Lewin JS, Duerk JL, et al. MR arthrography of the shoulder: rethinking traditional imaging procedures to meet the technical requirements of MR imaging guidance. AJR Am J Roentgenol 1997; 169:1453–7.

[6] Valls R, Melloni P. Sonographic guidance of needle position for MR arthrography of the shoulder. AJR Am J Roentgenol 1997;169:845–7.

[7] Trattnig S, Breitenseher M, Rand T, et al. MR imaging-guided MR arthrography of the shoulder: clinical experience on a conventional closed high-field system. AJR Am J Roentgenol 1999;172:1572–4.

[8] Grainger AJ, Elliott JM, Campbell RS, et al. Direct MR arthrography: a review of current use. Clin Radiol 2000;55:163–76.

[9] Steinbach LS, Palmer WE, Schweitzer ME. Special focus session. MR arthrography. Radiographics 2002; 22:1223–46.

[10] Wagner SC, Schweitzer ME, Weishaupt D. Temporal behavior of intraarticular gadolinium. J Comput Assist Tomogr 2001;25:661–70.

[11] Palmer WE, Caslowitz PL. Anterior shoulder instability: diagnostic criteria determined from prospective analysis of 121 MR arthrograms. Radiology 1995; 197:819–25.

[12] Palmer WE, Brown JH, Rosenthal DI. Rotator cuff: evaluation with fat-suppressed MR arthrography. Radiology 1993;188:683–7.

[13] Chung CB, Corrente L, Resnick D. MR arthrography of the shoulder. Magn Reson Imaging Clin N Am 2004;12:25–38.

[14] Brenner ML, Morrison WB, Carrino JA, et al. Direct MR arthrography of the shoulder: is exercise prior to imaging beneficial or detrimental? Radiology 2000; 215:491–6.

[15] Rafii M, Minkoff J. Advanced arthrography of the shoulder with CT and MR imaging. Radiol Clin North Aml 1998;36:609–33.

[16] Sciulli RL, Boutin RD, Brown RR, et al. Evaluation of the postoperative meniscus of the knee: a study comparing conventional arthrography, conventional MR imaging, MR arthrography with iodinated contrast material, and MR arthrography with gadolinium-based contrast material. Skeletal Radiol 1999;28:508–14.

[17] Lim PS, Schweitzer ME, Bhatia M, et al. Repeat tear of postoperative meniscus: potential MR imaging signs. Radiology 1999;210:183–8.

[18] Bronstein R, Kirk P, Hurley J. The usefulness of MRI in evaluating menisci after meniscus repair. Orthopedics 1992;15:149–52.

[19] Schimmer RC, Brulhart KB, Duff C, et al. Arthroscopic partial meniscectomy: a 12-year follow-up and two-step evaluation of the long-term course. Arthroscopy 1998;14:136–42.

[20] van Trommel MF, Simonian PT, Potter HG, et al. Different regional healing rates with the outside-in technique for meniscal repair. Am J Sports Med 1998; 26:446–52.

[21] Whitman TL, Diduch DR. Transient posterior knee

pain with the meniscal arrow. Arthroscopy 1998;14: 762–3.

[22] Smith DK, Totty WG. The knee after partial meniscectomy: MR imaging features. Radiology 1990;176: 141–4.

[23] Applegate GR, Flannigan BD, Tolin BS, et al. MR diagnosis of recurrent tears in the knee: value of intraarticular contrast material. AJR Am J Roentgenol 1993;161:821–5.

[24] Deutsch AL, Mink JH, Fox JM, et al. Peripheral meniscal tears: MR findings after conservative treatment or arthroscopic repair. Radiology 1990;176: 485–8.

[25] Farley TE, Howell SM, Love KF, et al. Meniscal tears: MR and arthrographic findings after arthroscopic repair. Radiology 1991;180:517–22.

[26] White LM, Schweitzer ME, Weishaupt D, et al. Diagnosis of recurrent meniscal tears: prospective evaluation of conventional MR imaging, indirect MR arthrography, and direct MR arthrography. Radiology 2002;222:421–9.

[27] Magee T, Shapiro M, Rodriguez J, et al. MR arthrography of postoperative knee: for which patients is it useful? Radiology 2003;229:159–63.

[28] Horton LK, Jacobson JA, Lin J, et al. MR imaging of anterior cruciate ligament reconstruction graft. AJR Am J Roentgenol 2000;175:1091–7.

[29] McCauley TR, Elfar A, Moore A, et al. MR arthrography of anterior cruciate ligament reconstruction grafts. AJR Am J Roentgenol 2003;181:1217–23.

[30] Rak KM, Gillogly SD, Schaefer RA, et al. Anterior cruciate ligament reconstruction: evaluation with MR imaging. Radiology 1991;178:553–6.

[31] Irizarry JM, Recht MP. MR imaging of the knee ligaments and the postoperative knee. Radiol Clin North Am 1997;35:45–76.

[32] Papakonstantinou O, Chung CB, Chanchairujira K, et al. Complications of anterior cruciate ligament reconstruction: MR imaging. Eur Radiol 2003;13: 1106–17.

[33] May DA, Snearly WN, Bents R, et al. MR imaging findings in anterior cruciate ligament reconstruction: evaluation of notchplasty. AJR Am J Roentgenol 1997; 169:217–22.

[34] McMahon PJ, Dettling JR, Yocum LA, et al. The cyclops lesion: a cause of diminished knee extension after rupture of the anterior cruciate ligament. Arthroscopy 1999;15:757–61.

[35] Jackson DW, Schaefer RK. Cyclops syndrome: loss of extension following intra-articular anterior cruciate ligament reconstruction. Arthroscopy 1990;6:171–8.

[36] O'Connor MA, Palaniappan M, Khan N, et al. Osteochondritis dissecans of the knee in children: a comparison of MRI and arthroscopic findings. J Bone Joint Surg Br 2002;84:258–62.

[37] Brittberg M, Winalski CS. Evaluation of cartilage injuries and repair. J Bone Joint Surg Am 2003; 85(Suppl 2):58–69.

[38] Bohndorf K. Osteochondritis (osteochondrosis) disse-

cans: a review and new MRI classification. Eur Radiol 1998;8:103–12.

[39] De Smet AA, Fisher DR, Graf BK, et al. Osteochondritis dissecans of the knee: value of MR imaging in determining lesion stability and the presence of articular cartilage defects. AJR Am J Roentgenol 1990; 155:549–53.

[40] De Smet AA, Ilahi OA, Graf BK. Reassessment of the MR criteria for stability of osteochondritis dissecans in the knee and ankle. Skeletal Radiol 1996;25: 159–63.

[41] Kramer J, Stiglbauer R, Engel A, et al. MR contrast arthrography (MRA) in osteochondrosis dissecans. J Comput Assist Tomogr 1992;16:254–60.

[42] Kramer J, Recht MP, Imhof H, et al. Postcontrast MR arthrography in assessment of cartilage lesions. J Comput Assist Tomogr 1994;18:218–24.

[43] Gagliardi JA, Chung EM, Chandnani VP, et al. Detection and staging of chondromalacia patellae: relative efficacies of conventional MR imaging, MR arthrography, and CT arthrography. AJR Am J Roentgenol 1994;163:629–36.

[44] Chandnani VP, Ho C, Chu P, et al. Knee hyaline cartilage evaluated with MR imaging: a cadaveric study involving multiple imaging sequences and intraarticular injection of gadolinium and saline solution. Radiology 1991;178:557–61.

[45] Palmer WE. MR arthrography: is it worthwhile? Top Magn Reson Imaging 1996;8:24–43.

[46] Woertler K, Buerger H, Moeller J, et al. Patellar articular cartilage lesions: in vitro MR imaging evaluation after placement in gadopentetate dimeglumine solution. Radiology 2004;230:768–73.

[47] Vande Berg BC, Lecouvet FE, Poilvache P, et al. Assessment of knee cartilage in cadavers with dual-detector spiral CT arthrography and MR imaging. Radiology 2002;222:430–6.

[48] Vande Berg BC, Lecouvet FE, Maldague B, et al. MR appearance of cartilage defects of the knee: preliminary results of a spiral CT arthrography-guided analysis. Eur Radiol 2004;14:208–14.

[49] Bashir A, Gray ML, Burstein D. Gd-DTPA2- as a measure of cartilage degradation. Magn Reson Med 1996;36:665–73.

[50] Bashir A, Gray ML, Boutin RD, et al. Glycosaminoglycan in articular cartilage: in vivo assessment with delayed Gd(DTPA)(2-)-enhanced MR imaging. Radiology 1997;205:551–8.

[51] Bashir A, Gray ML, Hartke J, et al. Nondestructive imaging of human cartilage glycosaminoglycan concentration by MRI. Magn Reson Med 1999;41: 857–65.

[52] Attarian DE, Guilak F. Observations on the growth of loose bodies in joints. Arthroscopy 2002;18:930–4.

[53] Brossmann J, Preidler KW, Daenen B, et al. Imaging of osseous and cartilaginous intraarticular bodies in the knee: comparison of MR imaging and MR arthrography with CT and CT arthrography in cadavers. Radiology 1996;200:509–17.

[54] Flanigan DP, Burnham SJ, Goodreau JJ, et al. Summary of cases of adventitial cystic disease of the popliteal artery. Ann Surg 1979;189:165–75.

[55] Ricci P, Panzetti C, Mastantuono M, et al. Cross-sectional imaging in a case of adventitial cystic disease of the popliteal artery. Cardiovasc Intervent Radiol 1999;22:71–4.

[56] Unno N, Kaneko H, Uchiyama T, et al. Cystic adventitial disease of the popliteal artery: elongation into the media of the popliteal artery and communication with the knee joint capsule: report of a case. Surg Today 2000;30:1026–9.

[57] Deutsch AL, Hyde J, Miller SM, et al. Cystic adventitial degeneration of the popliteal artery: CT demonstration and directed percutaneous therapy. AJR Am J Roentgenol 1985;145:117–8.

ELSEVIER
SAUNDERS

Radiol Clin N Am 43 (2005) 747 – 759

RADIOLOGIC
CLINICS
of North America

MR Imaging of the Diabetic Foot: Diagnostic Challenges

Deep S. Chatha, MD*, Patricia M. Cunningham, MD,
Mark E. Schweitzer, MD

Department of Radiology, Hospital for Joint Diseases Orthopaedic Institute, 301 East 17th Street, New York, NY 10003, USA

Pedal complications of diabetes develop in approximately 15% of diabetics over their lifetime, with approximately one fifth of diabetic patients being hospitalized at some point for pedal infections [1–4]. With approximately 15 million people affected by diabetes in the United States alone, this translates to almost 2.3 million patients who experience diabetic foot complications [5].

This problem is of key interest to the radiologist, because they are often consulted in the diagnosis and management of pedal diabetic complications. The societal costs are impressive with the cost of treating a single foot ulcer estimated at $28,000 over a 2-year period [6,7]. Further costs related to lower-extremity amputation in the diabetic population is estimated at over $1 billion with approximately 50,000 lower extremity amputations performed each year in the United States [8].

Amputation further predisposes to contralateral limb complications because of subsequent shifting of weight-bearing forces. Within 2 years of amputation, the contralateral lower extremity has a 50% incidence of serious complications with a risk of 50% to 66% of contralateral amputation within 5 years [9,10].

Pedal complications of diabetes have long presented a challenge for the clinician and radiologist. The foot is distinctly susceptible to the manifestations of diabetes because of numerous factors. The peripheral neuropathy, both motor and sensory, that results from diabetes subsequently leads to repetitive unrecognized microtrauma, and the peripheral autonomic neuropathy results in dry, cracked skin [11]. This may manifest initially as callus formation, and subsequently proceed to soft tissue injury and ulceration, osteomyelitis and abscess formation, and neuropathic joint [12,13]. The peripheral vascular disease results in poor healing and response to infection and chronic ischemia. In addition, several other metabolic derangements and hyperglycemia result in impaired immunologic function and wound healing. Debridement and amputation are often necessary [14,15].

The clinical manifestations of acute diabetic foot infection are often difficult to distinguish from changes related to neuroarthropathy [16]. Probing the ulcer bed to underlying bone is highly specific and easy to perform; however, it has very low sensitivity [17–19].

Imaging plays a key role in the identification of soft tissue, bony, and articular complications. The distinction is critical because the management differs significantly between these two entities. CT scan and plain films offer useful bony anatomic information. Their soft tissue detail is lacking, however, and their sensitivity and specificity for determining infection is low, especially in the early stages of infection [20–22]. Bone scan provides poor anatomic detail and often vague localization [23–26]. MR imaging is relied on primarily as the imaging tool for pedal complications related to diabetes [27–32].

MR imaging protocol for diabetic foot

Imaging of diabetic foot complications necessitates a good clinical history and often a quick examination by the performing radiologist to optimize the quality of the study. The location of the ulcer

---

* Corresponding author.
E-mail address: cathad01@med.nyu.edu (D.S. Chatha).

Table 1

MR imaging parameters for forefoot examination of the diabetic foot

Sequence	FOV	Matrix	TR	TE	T1	Bandwidth	ETL
Short-axis T1 non–fat sat	12	256 × 192	400–800	Min	—	16	—
Short-axis T2 TSE fat sat	12	256 × 256	2000–6000	60–70	—	16	16
Sagittal STIR	14	256 × 192	>2000	20–40	150	16	16
Short-axis VIBE 3DGRE fat sat pre- and postcontrast	12	256 × 192	200	4	—	16	—
Sagittal VIBE 3DGRE fat sat postcontrast	12	256 × 192	200	4	—	16	—

Short-axis plane is consistent, however, the corresponding longitudinal plane should be tangent to the area of interest (ie, if the ulcer is lateral or medial, a long axis view would be more valuable than a sagittal). If the MR imaging unit has high gradients, then a third postgadolinium administration plane can be added without adding too much time. FMPSPGR/T1 Fat Sat/VIBE 3DGRE sequences are all adequate postgadolinium sequences, however, T1 fat sat is far more time costly.

Abbreviations: ETL, echo train length; fat sat, saturated fat; FOV, field of view; TE, echo time; TR, repetition time; TSE, turbo spin echo; 3DGRE, three-dimensional gradient recalled.

should be clearly identified and the examination then tailored to that location. The entire foot should not be imaged when there is a specific location in question. Nor should both feet be imaged simultaneously, except in rare circumstances.

The forefoot is unique in its orientation, much like the hand, and the imaging planes required to fully evaluate this region are also unique. The foot and ankle require completely different imaging parameters.

If the forefoot and toes are the focus of the examination, then a small surface coil is ideal (either 3 or 5 in), or if the foot is not swollen, the wrist coil can be used. The plane perpendicular to the toes (short-axis view) is the key plane in which to evaluate ulcers and their relationship to underlying osseous structures (Table 1). A T1-weighted image is used to evaluate marrow changes and subcutaneous fat, and a fat-suppressed T2-weighted image, usually a STIR, is ideal in the short axis to evaluate for edema within the bone, soft tissues, and surrounding tendinous structures [33,34]. Sagittal views and a plane parallel to the toes (long-axis view) are ideal for imaging the metatarsophalangeal and interphalangeal joints in the evaluation of septic arthritis. If there are toe defor-

mities, the short-axis images may be misleading. It is usually possible, however, to "read through" this problem.

The midfoot and hindfoot should be imaged with an extremity coil, such as the chimney-type knee coil. The sagittal plane is ideal for evaluation of midfoot neuropathic involvement, the plantar surface, and the posterior calcaneus. A T1-weighted and a fat-suppressed fluid/T2-weighted image are both required in this plane. Axial and coronal planes are useful in the evaluation of the malleoli and the surrounding tendons (Table 2).

Contrast administration is extremely useful in the evaluation of diabetic pedal complications. Bland subcutaneous edema does not enhance following contrast administration; however, cellulitis often demonstrates patchy enhancement. Contrast also aids in the detection of sinus tracts and abscesses. Detection of osteomyelitis is also facilitated by contrast administration. It is imperative, however, that fat-saturated T1-type images be obtained both before and after contrast administration, in identical planes, to fully evaluate regions of abnormal enhancement. Most of the time these are not obtained with conventional

Table 2

MR imaging parameters for hind- and midfoot examination of the diabetic foot

Sequence	FOV	Matrix	TR	TE	T1	Bandwidth	ETL
Sagittal T1 TSE non–fat sat	16	256 × 192	400–800	Min	—	16	—
Sagittal STIR	18	256 × 192	>2000	20–40	150	16	16
Axial T2 fat sat	14	256 × 256	2000–6000	70	—	16	16
Sagittal VIBE 3DGRE fat sat pre- and postcontrast	16	256 × 192	200	4	—	16	—
Axial VIBE 3DGRE fat sat post contrast	14	256 × 256	200	4	—	16	—

Coronal imaging plane can be used if ulcers reside on the medial or lateral soft tissues, however, axial images will also cover this region adequately. If the MR imaging unit has high gradients, then a third postgadolinium administration plane can be added without adding too much time.

Abbreviations: ETL, echo train length; fat sat, saturated fat; FOV, field of view; STIR, short-T1 inversion recovery; TE, echo time; TR, repetition time; TSE, turbo spin echo; 3DGRE, three-dimensional gradient recalled.

spin echo technique, but more often with various types of two-dimensional and volume-acquired turbo gradient echo sequences (fast multislice spoiled gradient [FMPSPGR], volumetric interpolated breath-hold examination [VIBE], and so forth). It should be noted that the distinction is not whether an abnormality enhances, but how it enhances, which leads to the differentiation of neuropathic changes from infection, as discussed later.

Many of the diabetic feet that are imaged have previously undergone surgery for debridement, arthrodesis, or partial foot-sparing surgery. Surgical changes and areas of magnetic susceptibility can obscure regions of interest and contrast is very useful in this setting. Several metal suppression techniques may also be useful to decrease metallic artifact and increase conspicuity of underlying abnormalities. In these situations avoid using gradient echo technique and substitute STIR images for frequency-selective fat-suppressed images.

Diabetic foot infection

Callous or ulceration

One of the more common pedal complications of diabetes is that of callous formation and ulceration. Skin callous formation is often the result of abnormal biomechanics in the diabetic population

secondary to the peripheral neuropathy complicating this disease. In addition, abnormal areas of friction and poor-fitting footwear contribute further to callous formation.

The autonomic dysfunction that results from chronic diabetes also plays a role, because it contributes to dry callous from lack of perspiration. The lack of moisture precipitates skin cracking, which is further complicated by direct inoculation of infection [35].

The location of diabetic foot callous parallels the sites of skin pressure. In the nonneuropathic foot this occurs under the metatarsal heads, usually the second. If a hallux valgus is present, which is frequent in all neuropaths regardless of stage, callous occurs medial to the first metatarsal. In neuropathic feet with midfoot dominance and a rocker-bottom foot, callous has been shown to be most prominent under the cuboid. In most other neuropathic feet the callous is most prominent at the heads of the first and second metatarsal bones, at the fifth metatarsophalangeal joint, and on the plantar surface of the posterior calcaneus.

Callus is often appreciated on MR imaging and demonstrates low signal intensity on T1-weighted images. On T2-weighted images the callus generally appears low to intermediate signal. The appearance on MR imaging may sometimes be confusing, especially after contrast administration where there can be significant enhancement, in which callous may mimic focal soft tissue infection. The location of the callous and the lack of adjacent soft tissue changes, however, should aid in the diagnosis (Fig. 1).

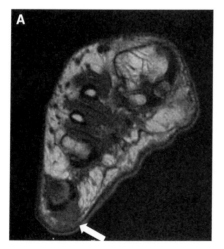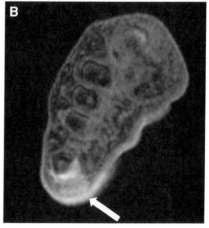

Fig. 1. Callous signal characteristics on MR imaging. (*A*) Short-axis T1-weighted image demonstrates focal area of low signal soft tissue within the plantar subcutaneous fat at the fifth metatarsal head (*arrow*). (*B*) Short-axis T1 fat-saturated image post–gadolinium administration demonstrates intense enhancement of the callous (*arrow*). Location of the callous and the lack of adjacent soft tissue changes should lead one away from misdiagnosing this as soft tissue infection.

The altered biomechanics and neuropathy result in callus hypertrophy. This is not, however, mechanically protective callus. The persistent weight-bearing and microtrauma in these regions results in callous breakdown. The result is focal ulceration. This focal ulceration, often combined with inoculation and decreased diabetic immune function, facilitates subsequent infection, often polymicrobial. These areas of ulceration have been shown to correspond with the regions related to persistent microtrauma and increased pressure during ambulation [36]. Callus breakdown with ulceration has been shown to be most prevalent at the first and fifth metatarsal heads, on the plantar surface of the second and third metatarsal heads, and on the dorsal surface of the toes. In the hindfoot, ulceration is most commonly seen at the posterior aspect of the calcaneus and at the medial and lateral malleoli.

On MR imaging ulceration is commonly identified as a defect of the overlying soft tissues, often adjacent to the bony prominence with granulation tissue seen at the ulcer base. The region of granulation tissue is low signal on T1-weighted images and high signal (or intermediate) on T2-weighted images. Following contrast administration there is intense enhancement at the ulcer base and often "tramtrack" peripheral enhancement at the developing sinus track (Fig. 2) [33,37].

Cellulitis and abscess

Although many ulcers are treated with antimicrobial therapy and if ischemic, debridement, many infected ulcers progress to more severe soft tissue infection, such as sinus tracks, severe cellulitis, and abscess formation. Cellulitis can often clinically be suggested on the basis of an erythematous, swollen, and warm lower extremity. This, unfortunately, is also the clinical presentation of acute neuropathic disease. The lower extremity and foot in noninfected, non-neuropathic is often edematous on MR imaging, secondary to soft tissue ischemia and venous insufficiency in diabetics. In both this "disease caused" benign edema and infectious cellulitis, the overlying skin may be mildly thickened and the subcutaneous fat reticulated on T1-weighted images. Both entities may also show increased signal on T2-weighted images. The administration of intravenous contrast proves very useful because regions of cellulitis demonstrate ill-defined enhancement. Diabetic ischemia does peculiar things to soft tissues, however, often shown by patchy irregular enhancement.

Focal abscess formation is not infrequent in the diabetic population, with reports estimating that be-

tween 10% and 50% of all cases that have pedal osteomyelitis have concomitant soft tissue abscesses [38,39]. As a general rule, abscesses are much more common in childhood musculoskeletal infections than adult ones. Among adults, however, diabetics have a not insignificant incidence.

Many of these intramuscular and intermuscular abscesses are quite small. There is invariable adjacent soft tissue infection with edema and rim enhancement [40]. Soft tissue abscesses demonstrate low to intermediate signal on T1-weighted images and high signal on T2-weighted images. There is substantial rim enhancement following intravenous contrast administration. Abscesses are significantly more common following inoculatory trauma (Fig. 3).

Abscesses, when present in the subcutaneous tissues, are quite easy to miss and are often fairly large. In this situation they may only show moderate reticulation of the subcutaneous fat.

Foreign body granuloma

The sensory neuropathy that accompanies diabetes may also manifest as inability to identify foreign bodies that have entered the superficial or deep soft tissues. These may become chronically imbedded and incite an inflammatory reaction without a frank abscess. There may be no accompanying soft tissue ulceration at the site of entry, and usually because of the sensory neuropathy, no history of trauma. The inflammatory changes are seen as high signal on T2-weighted images and low signal on T1-weighted images. Not infrequently, there may be little or no T2 signal changes, although marked enhancement is usually seen after contrast administration (Fig. 4).

The foreign body itself may be difficult to identify because they are usually quite small; however, an area of magnetic susceptibility or signal void should be sought within the granulomatous response and plain film or even CT correlation may also be useful to identify the inciting agent. It is not that surprising to suspect a body on MR imaging, yet not definitively visualize the body.

Septic arthritis

The interphalangeal and metatarsophalangeal joints are most commonly involved in septic arthritis, because of their superficial location and proximity to areas of ulceration [41,42]. Direct inoculation from the ulcer is the most common mode of infection in the diabetic population.

A joint effusion, often complex, is present on MR images of a septic joint [41]. Joint fluid is

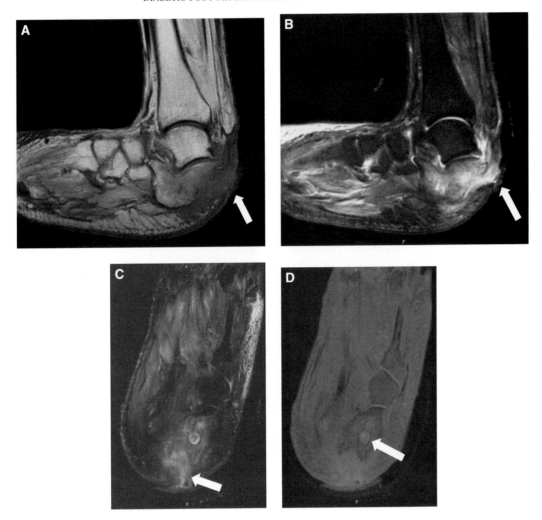

Fig. 2. Calcaneal ulcer with sinus tract and associated calcaneal osteomyelitis with intraosseous abscess. (*A*) Sagittal T1-weighted image of hindfoot shows large area of ulceration and soft tissue infiltration along the posterior aspect of the calcaneus with decreased signal (edema) in the adjacent bone marrow (*arrow*). Changes related to longstanding diabetic neuroarthropathy are also present with deformity of the midfoot and fatty infiltration of the intrinsic muscles. (*B, C*) Sagittal and axial T2 fat-saturated images of the hindfoot demonstrate significant bone marrow edema involving the calcaneus and surrounding soft tissue edematous changes. A focal sinus tract with a typical "tram-track" appearance is visualized in the soft tissues posteriorly (*arrow*). (*D*) Axial T1 fat-saturated image shows the subcutaneous soft tissue changes related to the ulcer bed and an intraosseous abscess in the posterior calcaneus (*arrow*). Abscesses in adult osteomyelitis are rare. They are somewhat more common in diabetics but tend, like in this case, to be small.

not synonymous with an effusion. Each joint has physiologic fluid. This becomes a diagnostic dilemma most often in the first metatarsophalangeal joint. Distention dorsally in this location is the best sign of an effusion. There may also be synovial or capsular distention at the affected joint. After contrast administration there is usually intense enhancement (Fig. 5).

The adjacent bony surfaces may demonstrate edematous changes, seen as increased T2 signal, which can be seen in either septic arthritis or osteomyelitis. In septic arthritis, the corresponding T1-weighted images should not demonstrate overt decreased signal, otherwise, osteomyelitis should be suspected. The most specific findings of septic arthritis are bony erosions, bone marrow edema, and carti-

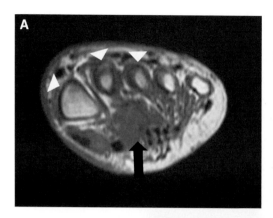
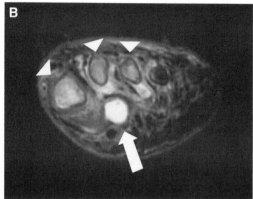
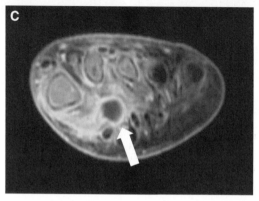

Fig. 3. Soft tissue abscess. (*A*) Short-axis T1-weighted image at the level of the mid-metatarsal region with focal low signal intensity mass between the second and third metatarsals, deep to the flexor tendons (*arrow*). The marrow signal of the first, second, and third metatarsals at this level demonstrate very mild decreased signal (*arrowheads*) compared with the fourth and fifth metatarsals. (*B*) Short-axis T2 fat-saturated image at the same level as Fig. 3A shows the focal mass is of fluid signal intensity (*arrow*). In addition, there is soft tissue edema involving the dorsal and plantar surface of the foot medially, and increased signal within the first through third metatarsals indicating marrow edema (*arrowheads*). (*C*) Short-axis T1 fat-saturated image post–gadolinium administration demonstrates significant soft tissue enhancement and focal ring enhancement of the deep soft tissue abscess (*arrow*).

lage destruction [41]. An adjacent soft tissue defect or ulcer also elevates the suspicion considerably.

Differentiating osteomyelitis from reactive bone marrow edema

The use of MR imaging to evaluate for osteomyelitis has increased steadily over the past decade. Sensitivity and specificity with respect to osteomyelitis have rates of over 90% [37]. Both osteomyelitis and bone marrow edema demonstrate high signal on T2-weighted images and inversion recovery images; however, osteomyelitis also demonstrates low bone marrow signal on T1-weighted images, whereas reactive bone marrow edema shows fairly normal T1 signal. Following contrast administration there is marrow enhancement in the presence of osteomyelitis, but sometimes also with reactive edema (Fig. 6).

The single best way to diagnose osteomyelitis is to find the ulcer or sinus track and follow it down to bone. If the marrow is abnormal signal on a T1-weighted image at that location, there is osteomyelitis present [37,41,42]. Note should then be made of extent of disease. Osseous extent is best determined on T1-weighted images. In fact, extent can be overdiagnosed on T2 or enhanced images. Articular involvement and soft tissue extent is best evaluated on contrast-enhanced images.

Spread of infection

There are three fascial compartments of the foot. The medial one goes to the base of the first meta-

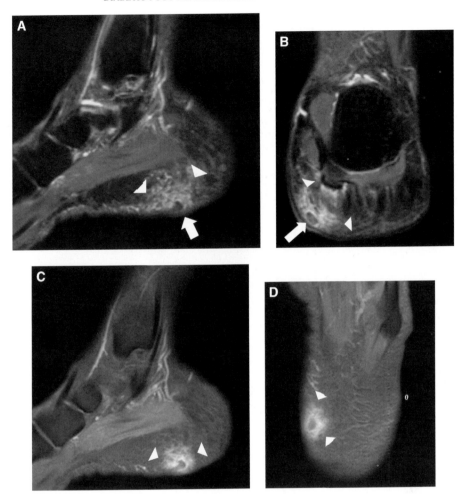

Fig. 4. Foreign body. Sagittal T2 fat-saturated image of the ankle (*A*) and a coronal T2 fat-saturated image at the level of the posterior calcaneus (*B*) demonstrate focal fatty infiltration and edema of the plantar soft tissues (*arrowheads*) associated with a small well-defined region of lower signal intensity centrally representing a foreign body (*arrow*). No magnetic susceptibility related to the foreign body is appreciated. Sagittal (*C*) and axial (*D*) T1 fat-saturated images post–gadolinium administration show nicely the focal soft tissue enhancement of the plantar surface (*arrowheads*) related to the foreign body. The site of the foreign body itself does not demonstrate significant enhancement.

tarsal. The lateral one extends to the base of the fifth metatarsal. The central one extends proximally into the mid and hindfoot and is actually contiguous with the calf muscles.

These compartments represent partial barriers to the spread of infection. That is, infection spreads proximally within them, rather than between them. The barriers are, however, imperfect and spread can occur, albeit less commonly, between compartments [43].

Joint spaces are, however, poor obstructions to the spread of infections, and periarticular osteomyelitis frequently spreads to the adjacent joint. Interestingly, tendons and their sheaths are not a common pathway for the spread of infections.

Several other points about tendons are important. First, although they are an infrequent route for proximal spread, their sheaths are not uncommonly infected [34]. Outside the hand, the most commonly infected tendons are the peroneals. That is because many ulcers occur at the lateral malleolus. Outside of the toes, where flexor digitorum and extensor digitorum infections, focally, are common, the next most common tendon infection is the Achilles. This is directly related to posterior ulceration.

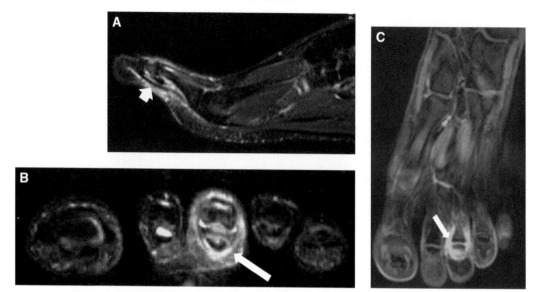

Fig. 5. Septic arthritis of interphalangeal joint. (*A, B*) Sagittal and short-axis STIR images of the forefoot with increased signal at the third proximal interphalangeal joint, flexor tendon sheath (*arrow*), and surrounding soft tissues. (*C*) Long-axis T1 fat-saturated post–gadolinium administration image demonstrating intense enhancement of the third proximal interphalangeal joint representing septic arthritis (*arrow*). A common location for ulceration is the dorsal interphalangeal joints, especially in patients who have more proximal neuropathic disease.

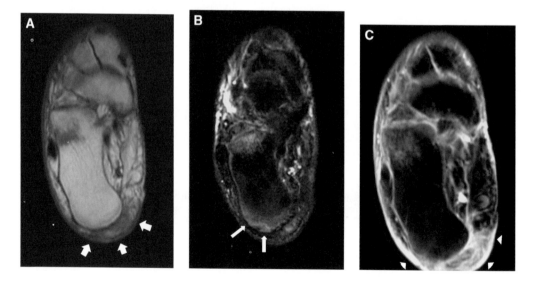

Fig. 6. Reactive bone marrow edema. (*A*) Axial T1-weighted image through the hindfoot demonstrates soft tissue infiltration within the posterior tissues (*arrows*), but normal bone marrow signal within the posterior calcaneus itself. (*B*) Axial T2 fat-saturated image at the same level shows moderate increased signal within the posterior calcaneus (*arrow*). (*C*) Axial T1 fat-saturated image post–gadolinium administration demonstrates moderate enhancement of the posterior soft tissues consistent with soft tissue inflammation and cellulitis; however, there is no enhancement of the underlying calcaneus. The T1 and postcontrast imaging features indicate that this is reactive bone marrow edema, rather than underlying osteomyelitis.

Most of these tendon infections are focal. Most soft tissue infections are local, close to the sinus track, or region of osteomyelitis. Distant spread, especially from medial or lateral compartment infections, is rare.

Care should be taken to assess for the degree of enhancement of the infected soft tissues. Devitalization is a common complication of diabetes. This is difficult to see using current imaging techniques but likely will be a significant indication for imaging in the next 2 years. Currently, a modestly abrupt cutoff of enhancement can be seen with devitalized soft tissues. In these patients, unfortunately, the usual MR imaging signs of osteomyelitis may be less reliable.

The presence of secondary signs is also useful in any differentiation of osteomyelitis from reactive bone marrow edema. An overlying cutaneous ulcer or sinus tract may be evident. Soft tissue ulceration over bony protuberances is often seen in diabetics and direct extension of infection is the primary mechanism of spread.

In addition, cortical interruption has been shown to have a high positive predictive value for osteomyelitis [37]. On T2-weighted sequences, periosteal reaction may be seen as circumferential high signal surrounding the cortex. This is seen to enhance avidly following contrast administration (Fig. 7). This can

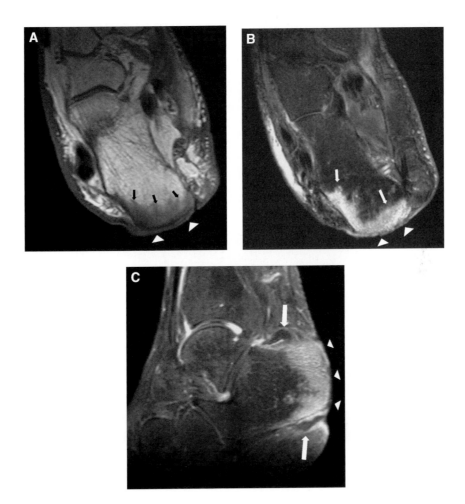

Fig. 7. Calcaneal osteomyelitis. (*A, B*) Axial T1-weighted image and T2 fat-saturated image of the hindfoot show extensive deep soft tissue ulceration (*arrowheads*) with associated focal decreased T1 bone marrow signal (*arrows*) in the posterior calcaneus corresponding to increased T2 signal, indicating osteomyelitis. (*C*) Sagittal T1 fat-saturated image post–gadolinium administration of the ankle showing enhancement within the posterior aspect of the calcaneus (*arrowheads*) with adjacent deep ulceration and soft tissue thinning. The posterior soft tissues are nearly absent. This is a common appearance in debilitated diabetics. Small fluid collections and abscesses are seen adjacent to the superior and inferior aspect of the calcaneus (*arrows*).

also be seen in neuropathic disease, especially about the proximal metatarsal shafts.

Neuroarthropathy

Differentiating diabetic neuroarthropathy versus osteomyelitis

Clinically and radiologically, neuroarthropathy can mimic osteomyelitis. The patient may present

with an erythematous and swollen foot in both situations. On MR imaging both entities may demonstrate subchondral bone marrow signal abnormalities and joint effusions, periosteal reaction, and soft tissue inflammatory changes.

Aid in the distinction between the two entities can be made on MR imaging mainly based on distribution of disease. Osteomyelitis is predisposed to occur at pressure points and areas of ulceration along bony protuberances [36,37]. Consequently, the most common locations for osteomyelitis are at the metatarsal

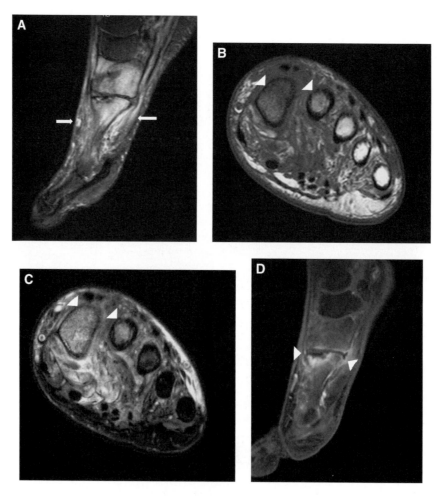

Fig. 8. Rapidly progressive neuroarthropathy in a diabetic patient. (*A*) Sagittal T2 fat-saturated image demonstrates diffuse soft tissue edema (*arrows*) involving the midfoot and significant bone marrow edema within the tarsal and metatarsal bones. (*B, C*) Short-axis T1 and T2 fat-saturated images show extensive soft tissue infiltration and edematous changes. Regions of bone marrow edema within the first and second metatarsals (*arrowheads*) correspond to regions of mildly decreased T1 marrow signal, which are suspicious for underlying osteomyelitis. (*D*) Sagittal T1 fat-saturated images after gadolinium administration demonstrate significant enhancement within the tarsometatarsal joints and somewhat less enhancement within the bones themselves. Although the signal characteristics are consistent with osteomyelitis, the lack of adjacent soft tissue defects or ulcers, sinus tracts, or abscess mitigates against osteomyelitis and supports the diagnosis of neuroarthropathy. Acute neuroarthropathy can clinically mimic osteomyelitis.

heads and at the interphalangeal joints in the forefoot and at the plantar aspect of the posterior calcaneus or distal fibula in the ankle and hindfoot.

Neuropathic changes tend to predominate in the midfoot. The only common location for infection or osteomyelitis in the midfoot is in the cuboid in severe, midfoot, neuropathic patients. MR imaging findings of neuroarthropathy in acute or rapidly progressive disease (Fig. 8) differ from those seen in more chronic or longstanding involvement (Fig. 9). In the acute setting, more prevalent bone marrow

edema and joint effusions are seen, leading these patients to have a more difficult MR imaging differentiation from acute infection or osteomyelitis.

Occasionally, the situation arises where infection is a concern in the patient who has neuropathic arthropathy. Recently the "ghost sign" was described as a potential useful observation to distinguish between the presence or absence of infection [31]. Areas of osteomyelitis on the background of a neuropathic joint may appear to be dissolved on T1-weighted images and subsequently may appear

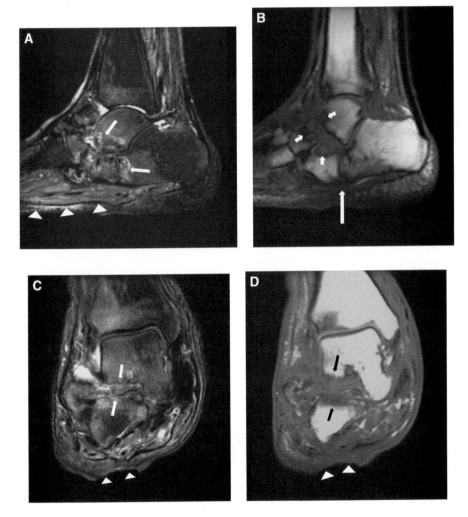

Fig. 9. Chronic neuroarthropathy. (*A*) Sagittal T2 fat-saturated image of the midfoot and hindfoot in a patient who has chronic neuroarthropathy demonstrating predominately midfoot destructive changes and bone marrow edema (*arrows*). Edema is also present within the plantar muscles, a finding often seen in chronic neuroarthropathy (*arrowheads*). (*B*) Sagittal T1-weighted image again shows the midfoot destructive changes (*short arrows*) and a rocker deformity at the calcaneocuboid articulation (*long arrow*). Coronal T2 fat-saturated image (*C*) and coronal T1-weighted image (*D*) through the hindfoot demonstrate hindfoot valgus secondary to destructive neuroarthropathy involving the subtalar joints (*arrows*) with associated bone marrow and soft tissue edema. A superficial plantar soft tissue ulcer is also present (*arrowheads*).

more regular on postcontrast T1-weighted images and on T2-weighted images. In addition, white cell scanning, either with indium or with in vivo labeling (Neutraspec), may be useful in this clinical circumstance.

Diabetic muscle infarction

Muscle ischemia and infarction is a well-documented although uncommon manifestation of diabetes. This complication has typically been described in the calf and thigh muscles of diabetics. Multiple sites of involvement are not uncommon and this can be seen bilaterally in up to one third of affected patients. It should be noted that bone infarcts are also somewhat increased in the diabetic population.

Clinically, the patient presents with pain and swelling over the involved area without a history of trauma [44,45]. Pain can be gradual or sudden in onset.

MR imaging of the affected region demonstrates enlarged, edematous muscles on T2-weighted and inversion-recovery images. The signal on T1-weighted images can be normal [46,47]. If the abnormal signal is localized to one muscle compartment and there are associated edematous changes, compartment syndrome must be considered as a secondary complication.

References

[1] Consensus Development Conference on Diabetic Foot Wound Care. 7–8 April 1999, Boston, Massachusetts. Diabetes Care 1999;22:1354–60.

[2] Palumbo PJ, Melton III LJ. Peripheral vascular disease and diabetes. In: Diabetes in America: diabetes data compiled 1984. NIH publication no. 85–1468. Washington: Government Printing Office; 1985. p. XV-1–XV-21.

[3] Mayfield JA, Reiber GE, Sanders LJ, et al. Preventive foot care in people with diabetes. Diabetes Care 1998; 21:2161–77.

[4] Smith DM, Weinberger M, Katz BP. A controlled trial to increase office visits and reduce hospitalizations of diabetic patients. J Gen Intern Med 1987;2:232–8.

[5] Boulton AJ, Vileikyte L. The diabetic foot: the scope of the problem. J Fam Pract 2000;49:S3–8.

[6] Ramsey SD, Newton K, Blough D, et al. Incidence, outcomes, and cost of foot ulcers in patients with diabetes. Diabetes Care 1999;22:382–7.

[7] Ragnarson Tennvall G, Apelqvist J. Prevention of diabetes related foot ulcers and amputations: a cost-utility analysis based on Markov model simulations. Diabetologia 2001;44:2077–87.

[8] Apelqvist J, Ragnarson Tennvall G, Persson U, et al. Diabetic foot ulcers in a multidisciplinary setting: an economic analysis of primary healing and healing with amputation. J Intern Med 1994;235:463–71.

[9] Goldner MG. The fate of the second leg in the diabetic amputee. Diabetes 1960;9:100–3.

[10] Hoar CS, Torres J. Evaluation of below the knee amputation in the treatment of diabetic gangrene. N Engl J Med 1962;266:440–3.

[11] Reiber GE, Vileikyte L, Boyko EJ, et al. Causal pathways for incident lower extremity ulcers in patients with diabetes from two settings. Diabetes Care 1999; 22:157–62.

[12] Joshi N, Caputo GM, Weitekamp MR, et al. Infections in patients with diabetes mellitus. N Engl J Med 1999; 341:1906–12.

[13] Karchmer AW, Gibbons GW. Foot infections in diabetes: evaluation and management. Curr Clin Top Infect Dis 1994;14:1–22.

[14] Lipsky BA, Berendt AR, Deery HG, et al. Diagnosis and treatment of diabetic foot infections. Clin Infect Dis 2004;39:885–910.

[15] van Baal JG. Surgical treatment of the infected diabetic foot. Clin Infect Dis 2004;39:S123–8.

[16] Berendt AR, Lipsky B. Is this bone infected or not? Differentiating neuro-osteoarthropathy from osteomyelitis in the diabetic foot. Curr Diab Rep 2004;4:424–9.

[17] Jeffcoate WJ, Lipsky BA. Controversies in diagnosing and managing osteomyelitis of the foot in diabetics. Clin Infect Dis 2004;39:S115–22.

[18] Boulton AJM, Kirsner RS, Vileikyte L. Neuropathic diabetic foot ulcers. N Engl J Med 2004;351:48–55.

[19] Grayson ML, Gibbons GW, Balogh K, et al. Probing to bone in infected pedal ulcers. JAMA 1995;273:721–3.

[20] David R, Barron BJ, Madewell JE. Osteomyelitis, acute and chronic. Radiol Clin North Am 1987;25: 1171–201.

[21] Schauwecker DS, Braunstein EM, Wheat LJ. Diagnostic imaging of osteomyelitis. Infect Dis Clin North Am 1990;4:441–63.

[22] Cavanagh PR, Young MJ, Adams JE, et al. Radiographic abnormalities in the feet of patients with diabetic neuropathy. Diabetes Care 1994;17:201–9.

[23] Kneenan AM, Nathaniel L, Tindel L, et al. Diagnosis of pedal osteomyelitis in diabetic patients using current techniques. Arch Intern Med 1989;149:2262–6.

[24] Balsells M, Viade J, Millan M, et al. Prevalence of osteomyelitis in non-healing diabetic foot ulcers: usefulness of radiologic and scintigraphic findings. Diabetes Res Clin Pract 1997;38:123–7.

[25] Becker W. Imaging osteomyelitis and the diabetic foot. Q J Nucl Med 1999;43:9–20.

[26] Lipman BT, Collier BD, Carrera GF, et al. Detection of osteomyelitis in the neuropathic foot: nuclear medicine, MRI and conventional radiography. Clin Nucl Med 1998;23:77–82.

[27] Cook TA, Rahim N, Simpson HC, et al. Magnetic resonance imaging in the management of diabetic foot infection. Br J Surg 1996;83:245–8.

[28] Croll SD, Nicholas GG, Osborne MA, et al. Role of magnetic resonance imaging in the diagnosis of osteomyelitis in diabetic foot infections. J Vasc Surg 1996;24:266–70.

[29] Yu JS. Diabetic foot and neuroarthropathy: magnetic resonance imaging evaluation. Top Magn Reson Imaging 1998;9:295–310.

[30] Boutin RD, Brossman J, Sartoris DJ, et al. Update on imaging of orthopedic infections. Orthop Clin North Am 1998;29:41–66.

[31] Schweitzer ME, Morrison WB. MR imaging of the diabetic foot. Radiol Clin North Am 2004;42:61–71.

[32] Gil HC, Morrison WB. MR imaging of diabetic foot infection. Semin Musculoskelet Radiol 2004;8: 189–98.

[33] Morrison WB, Schweitzer ME, Bock GW, et al. Diagnosis of osteomyelitis: utility of fat-suppressed contrast-enhanced MR imaging. Radiology 1993;189: 251–7.

[34] Ledermann HP, Morrison WB, Schweitzer ME, et al. Tendon involvement in pedal infection: MR analysis of frequency, distribution and spread of infection. American Journal of Radiology 2002;179:939–47.

[35] Lipsky BA, Pecoraro RE, Wheat LJ. The diabetic foot: soft tissue and bone infection. Infect Dis Clin North Am 1990;4:409–32.

[36] Ledermann HP, Morrison WB, Schweitzer ME. MR image analysis of pedal osteomyelitis: distribution, patterns of spread, and frequency of associated ulceration and septic arthritis. Radiology 2002;223: 747–55.

[37] Morrison WB, Schweitzer ME, Batte WG, et al. Osteomyelitis of the foot: relative importance of primary and secondary MR imaging signs. Radiology 1998;207:625–32.

[38] Beltran J, Campanini DS, Knight C, et al. The diabetic foot: magnetic resonance imaging evaluation. Skeletal Radiol 1990;19:37–41.

[39] Ledermann HP, Morrison WB, Schweitzer ME. Pedal abscesses in patients suspected of having pedal osteomyelitis: analysis with MR imaging. Radiology 2002;224:649–55.

[40] Ledermann HP, Schweitzer ME, Morrison WB. Nonenhancing tissue on MR imaging of pedal infection: characterization of necrotic tissue and associated limitations for diagnosis of osteomyelitis and abscess. American Journal of Radiology 2002;178:215–22.

[41] Karchensky M, Schweitzer ME, Morrison WB, et al. MRI findings of septic arthritis and associated osteomyelitis in adults. American Journal of Radiology 2004;182:119–22.

[42] Graif M, Schweitzer ME, Deely D, et al. The septic versus nonseptic inflamed joint: MRI characteristics. Skeletal Radiol 1999;28:616–20.

[43] Ledermann HP, Morrison WB, Schweitzer ME. Is soft tissue infection in pedal infection contained by fascial planes? MR analysis of compartmental involvement in 115 feet. American Journal of Radiology 2002;178: 605–12.

[44] Habib GS, Nashashibi M, Saliba W, et al. Diabetic muscle infarct: emphasis on pathogenesis. Clin Rheumatol 2003;22:450–1.

[45] Ly JQ, Yi EK, Beall DP. Diabetic muscle infarction. AJR Am J Roentgenol 2003;181:1216.

[46] Khoury NJ, El-Khoury GY, Kathol MH. MRI diagnosis of diabetic muscle infarction: report of two cases. Skeletal Radiol 1997;26:122–7.

[47] Chason DP, Flekenstein JL, Burns DK, et al. Diabetic muscle infarction: radiologic evaluation. Skeletal Radiol 1996;25:127–32.

ELSEVIER
SAUNDERS

Radiol Clin N Am 43 (2005) 761 – 770

RADIOLOGIC
CLINICS
of North America

Normal Variants and Frequent Marrow Alterations that Simulate Bone Marrow Lesions at MR Imaging

Bruno C. Vande Berg, MD, PhD*, Frédéric E. Lecouvet, MD, PhD,
Christine Galant, MD, Baudouin E. Maldague, MD, Jacques Malghem, MD

Saint Luc University Hospital, Université de Louvain, Hippocrate Avenue 10/2942, B-1200 Brussels, Belgium

MR imaging of the spine is routinely performed for the assessment of patients with spine-related symptoms and of patients with cancer. This article addresses normal variants and frequent alterations of the vertebral bone marrow that are encountered on MR imaging studies and can simulate lesions.

MR imaging appearance of the normal vertebral marrow

MR imaging appearance of the vertebral marrow merely depends on the relative proportion of hematopoietic cells and adipocytes within the medullary cavity of the vertebral bodies [1,2]. Most lesions and normal variants interfere with this medullary water-fat balance and are best depicted on T1-weighted spin echo (SE) MR images [2].

Normal vertebral marrow of the adult human shows intermediate signal intensity on both T1- and T2-weighted SE images (Figs. 1 and 2). As a rule, signal intensity of normal lumbar vertebral bodies on T1-weighted SE images must be higher than that of adjacent intervertebral disk in an adult patient [3]. In the thoracic spine, marrow signal intensity can be lower than that of disk because the disk can show higher signal intensity on images than in the lumbar

spine. In the pelvis, marrow signal intensity should be higher than that of adjacent normal muscles on T1-weighted SE images. On T2-weighted SE images, there is no internal standard with which marrow signal intensity can be compared. It is unreliable to assess the marrow status on T2-weighted SE images. On fat-saturated T2- or intermediate-weighted fast SE images, vertebral marrow signal intensity normally ranges from intermediate to moderately elevated. After intravenous injection of gadolinium-containing contrast material, enhancement of marrow signal intensity is barely visible at visual inspection on T1-weighted SE images (see Fig. 1). Signal enhancement of the intramedullary and perivertebral veins is visible on contrast-enhanced T1-weighted SE images. Signal enhancement can be more obvious on fat-saturated T1-weighted SE images or can be quantitatively assessed by performing dynamic MR imaging studies [4,5]. Usually, normal marrow signal intensity should not increase by more than 35% in adults above 35 years of age.

It is important to keep in mind that even a normal-looking T1-weighted SE image of the spine does not enable one to exclude marrow infiltration by abnormal plasmocytes or lymphocytes, probably in situations in which the water-fat balance is not sufficiently altered to become detectable.

Important interindividual variations in vertebral marrow appearance

There are important interindividual variations in vertebral marrow appearance among normal subjects.

---

* Corresponding author.
E-mail address: vandeberg@rdgn.ucl.ac.be
(B.C. Vande Berg).

0033-8389/05/$ – see front matter © 2005 Elsevier Inc. All rights reserved.
doi:10.1016/j.rcl.2005.01.007

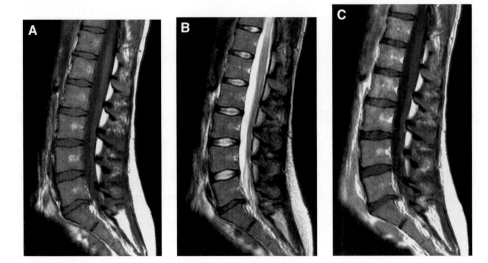

Fig. 1. Normal marrow. (*A*) Sagittal T1- and (*B*) T2-weighted spin echo (SE) of the lumbar spine of a 23-year-old woman show homogeneous marrow appearance with more fatty marrow around the vertebral veins. (*C*) On the gadolinium-enhanced sagittal T1-weighted image, moderate signal intensity enhancement can be seen, merely by noting the artifactual decrease in signal intensity of the intervertebral disk. Enhancement percentage was 70%, as determined on quantitative dynamic MR imaging study (not shown).

Vertebral marrow of young adults normally shows intermediate signal intensity on T1-weighted images. Other subjects of the same age range, however, can show unusual high marrow signal intensity with respect to their age. Conversely, some elderly subjects can show relatively low signal intensity marrow, instead of the expected age-related high signal intensity marrow on T1-weighted SE images. Although technical parameters may partially account for interindividual variability, interindividual variations in red marrow cellularity of the vertebral marrow probably play a role, whatever the causes.

Limited intervertebral variations in vertebral marrow appearance

There is limited variation in marrow appearance among vertebral bodies of the same subject, in contrast to marked interindividual variability in vertebral

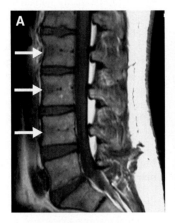
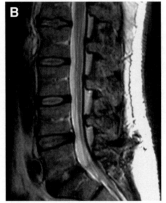

Fig. 2. Normal marrow. (*A*) Sagittal T1- and (*B*) T2-weighted SE of the lumbar spine of a 44-year-old man show moderate decrease in signal intensity in the anterior aspect of the vertebral bodies (*arrows*). Note that the same pattern of more cellular red marrow distribution is present in all vertebral bodies.

marrow appearance. Actually, several patterns of vertebral marrow appearance can be recognized that are systematically observed in all vertebral bodies of the same subject [6,7]. Red marrow is generally distributed in a homogeneous pattern within the vertebral body (see Fig. 1). Occasionally, red marrow is more cellular near the vertebral end plate, which is a metaphyseal equivalent, an area where the vasculature is generally more developed [8]. Red marrow can also be more cellular in the anterior aspects of the vertebral bodies (see Fig. 2). Finally, fatty marrow can become prominent around the vertebral basilar veins (see Figs. 1 and 2) [8]. As a rule, these variations in vertebral marrow MR imaging appearance should involve all vertebral bodies of the same subject in a similar manner (no or minor intervertebral variations).

Normal variants

Normal variants of vertebral marrow result from focal or diffuse changes in the amount of normal marrow components including fat and hematopoietic cells.

Islands of fatty marrow

At birth, vertebral marrow shows homogeneous low signal intensity. During growth, the proportion of marrow fat cells increases in a diffuse and homogeneous manner, a process called marrow conversion, which results in a progressive increase in marrow signal intensity on T1-weighted images with age [9]. During adulthood, conversion of red to yellow marrow continues at a lower pace and in a more heterogeneous manner than during growth. Foci of yellow marrow appear in the vertebral bodies (Fig. 3) [10]. Their frequency increases with age, but their number and size remain unaltered at short-term follow-up MR imaging study. As expected, these islands of fatty marrow show high signal intensity on T1-weighted SE images and low signal intensity on fat-saturated images. On non–fat-saturated intermediate or T2-weighted fast SE images, these areas show high signal intensity. They can be confused with significant marrow lesions. Analysis of the corresponding T1-weighted SE images or of fat-saturated images enables one to recognize their fatty composition, and their lack of clinical significance. CT images of these areas show normal trabecular and cortical bone. Bone scan shows normal or slightly decreased uptake in the corresponding area, if its size is large enough to enable their detection.

Islands of red marrow

Random variations in red marrow distribution occasionally do occur in addition to several patterns of vertebral red marrow distribution (see Fig. 3). Actually, cellularity of hematopoietic marrow can show spatial variations with the presence of islands

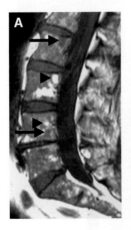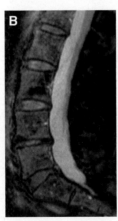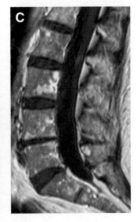

Fig. 3. Heterogeneous marrow. (*A*) Sagittal T1-weighted SE and (*B*) fat-saturated intermediate-weighted fast spin echo (FSE) of the lumbar spine of a 73-year-old woman with breast cancer show heterogeneous marrow appearance with islands of fatty marrow (*arrowheads*) and small nodules of more cellular marrow (*arrows*). Areas of more cellular marrow also show intermediate signal on fat-saturated image. (*C*) On the gadolinium-enhanced sagittal T1-weighted image, moderate signal intensity enhancement can be seen. This signal pattern remains compatible with normal heterogeneous marrow. A biopsy of the iliac crest was performed during breast surgery and showed hypercellular reactive hematopoietic marrow devoid of neoplastic cells.

of highly cellular hematopoietic marrow. These variations lead to the appearance of areas of more pronounced decrease in signal intensity than adjacent marrow on T1-weighted SE MR images. They occur in a nonpredictable manner, although they frequently involve the peripheral aspect of the vertebral bodies. Their margins are sharp if the marrow conversion process is advanced and fuzzier if the marrow conversion process is limited [11]. Occasionally, central areas of high signal intensity on T1-weighted images are present, which are an additional argument in favor of a normal variant. Presence of low-to-intermediate signal intensity on T2-weighted images, lack of evident signal enhancement on T1-weighted images after gadolinium injection, lack of trabecular bone changes on CT images, and lack of changes at follow-up MR imaging studies generally help to differentiate these benign heterogeneities from clinically relevant abnormalities [1].

Focal vertebral alterations that may simulate metastasis

Vertebral hemangioma

Vertebral hemangioma is a common vertebral lesion with a frequency of 12% in women and 9% in men (Fig. 4) [12]. Hemangiomas are multiple in about one third of cases [12]. They are generally asymptomatic. Histologically, they correspond to cavernous hemangiomas and contain dilated, blood-

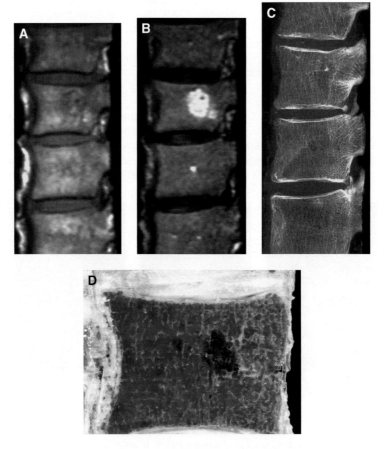

Fig. 4. Vertebral hemangioma. (*A*) Sagittal T1-weighted SE and (*B*) T2-weighted FSE of the lumbar spine of a cadaver specimen show an area of high signal intensity on T2-weighted image that shows almost normal signal intensity on T1-weighted SE image. There is a small punctate area of low signal within the lesion. (*C*) Radiograph of the specimen shows almost normal trabecular bone pattern with a small sclerotic island. (*D*) Photograph of the corresponding anatomic section shows a vertebral hemangioma with small punctuated area that corresponds to dilated vessels.

filled, vascular spaces lined by flat endothelial cells, set in a stroma containing large amounts of adipose tissue and no hematopoietic cells [13,14]. On T1-weighted SE images, signal intensity of asymptomatic vertebral hemangioma is higher to that of adjacent marrow (see Fig. 4) [15], although it can also be equivalent and not visible on T1-weighted images (Fig. 5). On T2-weighted SE images, its signal is consistently high (see Fig. 4). Presence of fat cells and dilated vessels with interstitial edema most likely accounts for its high signal intensity on T1- and T2-weighted images, respectively [16]. Frequently, punctuated or linear areas of low signal intensity are also seen on T1- and T2-weighted images, probably caused by the presence of thickened trabeculae. Signal enhancement of hemangioma after gadolinium injection is variable, depending on its appearance on T1-weighted images and the type of sequence that is obtained after contrast injection. Enhancement pattern can be homogeneous or peripheral.

Occasionally, asymptomatic vertebral hemangioma shows low signal intensity on T1-weighted images, with marked enhancement on postcontrast T1-weighted SE images (Fig. 6). These hemangiomas can be confused with significant marrow lesions. CT images generally show a rather specific trabecular

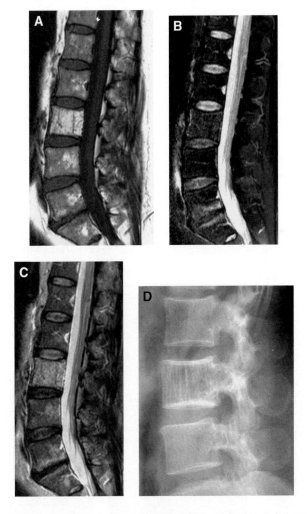

Fig. 5. Typical vertebral hemangioma. (*A*) Sagittal T1-weighted SE, (*B*) T2-weighted FSE, and (*C*) fat-saturated intermediate-weighted FSE images of the lumbar spine of a 54-year-old woman shows marrow change in the L3 vertebral body with high signal on T1, intermediate signal on T2, and subtle increased signal on fat-saturated intermediate-weighted FSE images suggestive of a vertebral hemangioma. (*D*) Lateral radiograph of the spine shows thickened vertical trabeculae without vertebral hypertrophy suggestive of vertebral hemangioma.

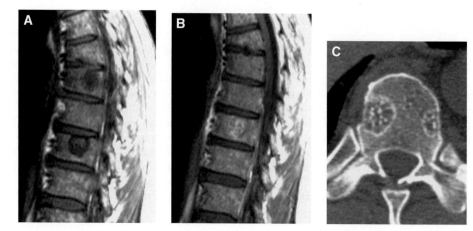

Fig. 6. Atypical vertebral hemangioma. (*A, B*) Precontrast and postcontrast sagittal T1-weighted SE images of the thoracic spine of a 69-year-old man show two areas of low signal intensity that enhance after contrast injection. (*C*) Axial CT image shows the typical appearance of a vertebral hemangioma. Rarely, hemangioma can show a nonspecific pattern of decreased signal on T1-weighted SE images.

bone pattern suggestive of vertebral hemangioma (see Fig. 6) [14,17], although small hemangiomas can remain occult of CT images.

The symptomatic vertebral hemangioma generally demonstrates low signal intensity on T1-weighted images with extraosseous component [14,17].

Vertebral enostosis: compact bone island

A compact bone island consists of lamellar cortical bone embedded within the trabecular network of the medullary cavity [18]. In a radiologic study of cadaver spine, their frequency was 14% and their size varied between 2 and 10 mm [19]. They frequently involve the periphery of the vertebral bodies and spare the central area [20]. Their signal intensity is very low on all sequences and adjacent marrow generally has a normal appearance (Fig. 7) [18]. Rarely, a peripheral high signal intensity rim surrounding a central low signal intensity area has been reported on STIR images of compact bone islands (Fig. 8) [21]. This pattern must be considered to be

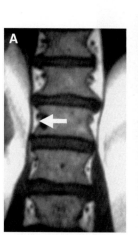
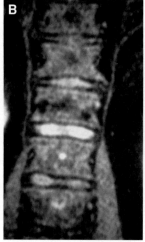
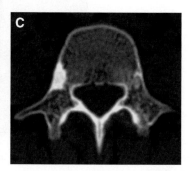

Fig. 7. Typical vertebral enostosis. (*A*) Coronal T1-weighted and (*B*) STIR images show an area of very low signal intensity in the right aspect of the vertebral body (*arrow*). (*C*) Axial CT image of that vertebral body shows a typical bone island, adjacent to the vertebral wall.

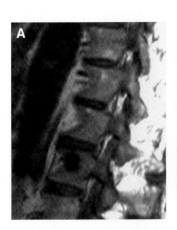
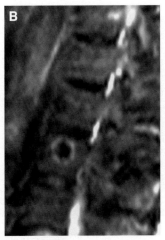
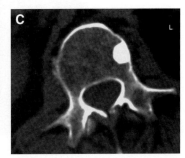

Fig. 8. Atypical enostosis. (*A*) Sagittal T1-weighted image of the lumbar spine shows an area of decreased signal intensity. (*B*) On the sagittal fat-saturated intermediate-weighted FSE image, the lesion shows a central area of low signal intensity and a peripheral rim of high signal intensity. (*C*) Axial CT image shows a vertebral enostosis. Bone scintigraphy (not shown) was normal. Rarely, vertebral enostosis may show a peripheral rim of high signal intensity on fat-saturated intermediate-weighted FSE images.

exceedingly rare and is more frequently suggestive of sclerotic metastases than an uncommon bone island.

Focal nodular hyperplasia of red marrow

Focal nodular hyperplasia of red marrow is the most extreme pattern of focal hypertrophy of the red marrow component (Fig. 9). It causes the presence of one or multiple nodules of decreased signal on T1-weighted images. It occurs relatively rarely in normal individuals but it is frequent in patients with regenerating hematopoietic marrow after marrow aplasia or in response to the administration of hematopoietic growth factors. Typically, the signal of these areas remains compatible with that of normal red marrow including moderate decrease in signal intensity on T1-weighted SE images, intermediate to low signal intensity of T2 fast SE images, moderate increase in signal intensity on fat-saturated intermediate- or T2-weighted fast SE images, and no or moderate enhancement after gadolinium injection. Most frequently, the nodule signal intensity is similar or slightly lower than that of the adjacent intervertebral disk [22]. Short-term follow-up MR imaging

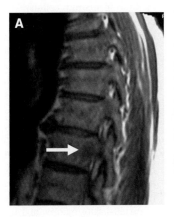
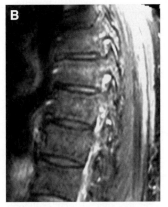

Fig. 9. Nodule of red marrow. (*A*) Sagittal T1-weighted and (*B*) fat-saturated intermediate-weighted FSE images of the thoracic spine of a 73-year-old man with lung cancer show an area of slight decrease in signal intensity (*arrow*). The signal of the lesion is similar to that of the intervertebral disk. Biopsy of another similar (larger) lesion in the same patient demonstrated normal but hypercellular marrow.

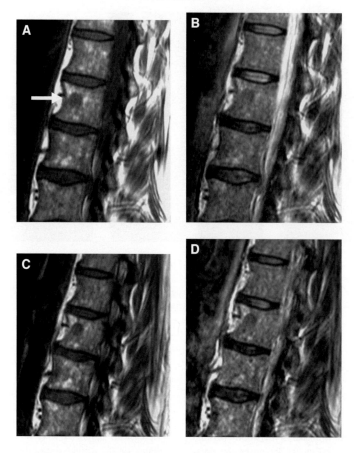

Fig. 10. (*A, B*) Sagittal T1- and T2-weighted SE images of the spine of a 56-year-old woman show an area of slight decrease in signal intensity (*arrow*). The signal intensity of the lesion is intermediate on T1- and T2-weighted SE images, and the lesion is compatible with an area of red marrow. Bone scintigraphy and CT scan were normal. (*C, D*) Corresponding images obtained 3 years later show no change in MR imaging appearance, increasing the likelihood of a nonsignificant marrow alteration.

study demonstrates no change in MR imaging appearance (Fig. 10). CT images and bone scan images should be normal. Experience with G6 PD fluorodeoxyglucose–positron emission tomography imaging is limited but these areas may show slight increased hypermetabolic uptake in comparison with adjacent red marrow [22].

Diffuse hematopoietic marrow hyperplasia

Diffuse hematopoietic marrow hyperplasia is defined by the presence of hypercellular hematopoietic marrow in the axial skeleton and by expansion of hematopoietic marrow in the appendicular skeleton (marrow reconversion). It occurs in response to numerous stimuli that trigger production of red marrow cells. Administration of hematopoietic growth factors during chemotherapy typically causes transient increase in marrow cellularity [23]. Several chronic disorders that are associated with anemia including hereditary hemoglobinopathies and chronic infection also cause red marrow expansion [24]. Diffuse red marrow hyperplasia is also observed in middle-aged obese women; in heavy smokers; and in subjects with intensive sports activities, such as long-distance running [25,26].

On T1-weighted SE images, hematopoietic marrow hyperplasia is associated with a marked decrease in signal intensity of vertebral marrow that becomes lower than that of adjacent disk (or gluteus muscles in the pelvis). Occasionally, hematopoietic marrow hyperplasia is heterogeneous because of the presence of residual fatty marrow and foci of red marrow. On T2-weighted SE images, vertebral signal intensity is low probably because red marrow hyperplasia is associated with an increase in the intracellular amount of iron. Intermediate signal intensity is observed on fat-saturated intermediate-weighted images. After intravenous gadolinium injection, signal intensity enhancement is moderate but can increase up to 80% on dynamic T1-weighted SE images. In the appendicular skeleton, expansion of red marrow in distal limbs can be observed along with nodules of regenerating red marrow that can simulate bone metastasis.

Diffuse hematopoietic marrow hyperplasia can be confused with diffuse marrow infiltration in the spine and with focal metastases in the limbs [27]. As a rule, marrow hyperplasia shows a signal similar to that of red marrow. The signal intensity pattern of diffuse marrow infiltration by neoplastic cells can be different to that of red marrow. There are cases, however, in which confusion is possible. Several techniques can be used to differentiate red marrow hyperplasia from red marrow infiltration by neoplastic cells. In- and out-phase gradient echo images, T1 relaxation time determination, hydrogen proton spectroscopy, dynamic contrast MR imaging studies, and diffusion-weighted images have all shown to be of some help but none has demonstrated definite conclusive results. Fluorodeoxyglucose positron emission tomography imaging has also shown limitation in this setting because diffuse increased uptake can be observed in red marrow hyperplasia as in neoplastic medullary infiltration [28,29]. Obtaining a blind iliac crest biopsy may remain the more accurate technique to address this occasionally difficult problem.

Summary

MR imaging appearance of normal marrow shows important variations among individuals because of focal or diffuse alteration in the amount of yellow and red marrow. Nodules of hematopoietic marrow remain difficult to characterize, mainly because no single noninvasive imaging modality is able to demonstrate specific features. Diffuse red marrow hyperplasia remains a diagnostic challenge but is easily accessible to blind iliac crest biopsy, if necessary. Vertebral enostosis and hemangioma show a spectrum of changes on MR images but other imaging modalities may contribute to their characterization.

References

[1] Vande Berg BC, Malghem J, Lecouvet FE, et al. Magnetic resonance imaging of the normal bone marrow. Skeletal Radiol 1998;27:471–83.

[2] Vogler JBI, Murphy WA. Bone marrow imaging. Radiology 1988;168:679–93.

[3] Carroll KW, Feller JF, Tirman PF. Useful internal standards for distinguishing infiltrative marrow pathology from hematopoietic marrow at MRI. J Magn Reson Imaging 1997;7:394–8.

[4] Montazel JL, Divine M, Lepage E, et al. Normal spinal bone marrow in adults: dynamic gadolinium-enhanced MR imaging. Radiology 2003;229:703–9.

[5] Chen WT, Shih TT, Chen RC, et al. Vertebral bone marrow perfusion evaluated with dynamic contrast-enhanced MR imaging: significance of aging and sex. Radiology 2001;220:213–8.

[6] De Bruyn PPH, Breen PC, Thomas TB. The microcirculation of the bone marrow. Anat Rec 1970;168:55–68.

[7] Weiss L. The structure of bone marrow functional interrelationships of vascular and hematopoietic compartments in experimental hemolytic anemia: an electron microscopic study. J Morphol 1965;117:467–538.

[8] Ricci C, Cova M, Kang YS, et al. Normal age-related patterns of cellular and fatty bone marrow distribution in the axial skeleton: MR imaging study [see comments]. Radiology 1990;177:83–8.

[9] Cristy M. Active bone marrow distribution as a function of age in humans. Phys Med Biol 1981;26:389–400.

[10] Hajek PC, Baker LL, Goobar JE, et al. Focal fat deposition in axial bone marrow: MR characteristics. Radiology 1987;162(1 Pt 1):245–9.

[11] Levine CD, Schweitzer ME, Ehrlich SM. Pelvic marrow in adults. Skeletal Radiol 1994;23:343–7.

[12] Schmorl G, Junghans H, Doin G. Clinique et radiologie de la colonne vertébrale normale et pathologique. In: Doin G, editor. Lésion du rachis osseux. Paris: Flammarion; 1956. p. 71–140.

[13] Murphey MD, Fairbain KJ, Parman L, et al. Muskuloskeletal angiomatous lesions: radiologic-pathologic correlation. Radiographics 1995;15:893–917.

[14] Wilner D. Radiology of bone tumors and allied disorders. In: Wilner D, editor. Benign vascular tumors and allied disorders of bone. Philadelphia: WB Saunders; 1982. p. 660–782.

[15] Ross JS, Masaryk TJ, Modic MT, et al. Vertebral hemangiomas: MR imaging. Radiology 1987;165:165–9.

[16] Baudrez V, Gallant C, Lecouvet FE, et al. Vertebral hemangioma: MR-histological correlation in autopsy specimens. Radiology 1999;213(P):245.

[17] Laredo JD, Reizine D, Bard M, et al. Vertebral hemangiomas: radiologic evaluation. Radiology 1986;161:183–9.

[18] Murphey MD, Andrews CL, Flemming DJ, et al. Primary tumors of the spine: radiologic-pathologic correlation. Radiographics 1996;16:1131–58.

[19] Kroon HM, Bloem JL, Holscher HC, et al. MR imaging of edema accompanying benign and malignant bone tumors. Skeletal Radiol 1994;23:261–9.

[20] Resnick D, Nemcek Jr AA, Haghighi P. Spinal enostoses (bone islands). Radiology 1983;147:373–6.

[21] Seymour R, Davies AM, Evans N, et al. Diagnostic problems with atypical bone islands. Br J Radiol 1997; 70:87–8.

[22] Bordalo-Rodrigues M, Galant C, Lonneux M, et al. Focal nodular hyperplasia of the hematopoietic marrow simulating vertebral metastasis on FDG positron emission tomography. AJR Am J Roentgenol 2003; 180:669–71.

[23] Altehoefer C, Bertz H, Ghanem NA, et al. Extent and time course of morphological changes of bone marrow induced by granulocyte-colony stimulating factor as assessed by magnetic resonance imaging of healthy blood stem cell donors. J Magn Reson Imaging 2001;14:141–6.

[24] Stabler A, Doma AB, Baur A, et al. Reactive bone marrow changes in infectious spondylitis: quantitative assessment with MR imaging. Radiology 2000;217: 863–8.

[25] Shellock FG, Morris E, Deutsch AL, et al. Hematopoietic bone marrow hyperplasia: high prevalence on MR images of the knee in asymptomatic marathon runners. AJR Am J Roentgenol 1992;158:335–8.

[26] Poulton TB, Murphy WD, Duerk JL, et al. Bone marrow reconversion in adults who are smokers: MR imaging findings. AJR Am J Roentgenol 1993;161: 1217–21.

[27] Ciray I, Lindman H, Astrom GK, et al. Effect of granulocyte colony-stimulating factor (G-CSF)-supported chemotherapy on MR imaging of normal red bone marrow in breast cancer patients with focal bone metastases. Acta Radiol 2003;44:472–84.

[28] Hollinger EF, Alibazoglu H, Ali A, et al. Hematopoietic cytokine-mediated FDG uptake simulates the appearance of diffuse metastatic disease on whole-body PET imaging. Clin Nucl Med 1998;23:93–8.

[29] Elmstrom RL, Tsai DE, Vergilio JA, et al. Enhanced marrow [18F]fluorodeoxyglucose uptake related to myeloid hyperplasia in Hodgkin's lymphoma can simulate lymphoma involvement in marrow. Clin Lymphoma 2004;5:62–4.

ELSEVIER
SAUNDERS

Radiol Clin N Am 43 (2005) 771 – 807

**RADIOLOGIC
CLINICS**
of North America

The Complementary Roles of MR Imaging and Ultrasound of Tendons

Ronald S. Adler, MD, PhD*, Kathleen C. Finzel, MD

*Department of Radiology and Imaging, Hospital for Special Surgery, Weill Medical College of Cornell University,
535 East 70th Street, New York, NY 10021, USA*

Most tendon pathology is imaged using MR imaging in the United States. In some institutions, however, ultrasound (US) has become either the principal method to assess tendon pathology or serves an important adjunctive role to MR imaging. Outside of the United States, US may be more widely used as the method of choice to evaluate tendon pathology. Although this may reflect the limited availability of MR imaging outside the United States, US has, nevertheless, been shown to be a reliable method to assess a large variety of musculoskeletal abnormalities.

A variety of studies comparing the efficacy of MR imaging and US have been performed, showing that both are capable of accurately characterizing tendon abnormalities [1–5]. These comparisons are most prevalent with regard to the rotator cuff. It is unfortunate that most comparisons of US and MR imaging put these modalities at odds with one another rather than emphasizing their complementary roles in assessing musculoskeletal pathology. There is clearly considerable overlap between these two modalities, where they may be used interchangeably to make a diagnosis. Examples include the Achilles tendon and rotator cuff. There are also situations, however, where one or another modality may be more appropriate to address a specific issue, or where the addition of another study is beneficial in making a specific di-

agnosis or impacts clinical outcome. An example is a twisting injury, where tendon or peripheral ligament damage is evident on US, but intrinsic ligamentous injury, marrow contusions, and occult cartilage injury require additional imaging with MR. Likewise, a small metallic screw impinging the deep surface of a tendon is best examined sonographically, because of the absence of interfering artifact, and US's dynamic capability.

This article demonstrates the complementary roles of these two modalities through a variety of clinical examples, based on experience working in a subspecialty hospital dedicated to orthopedic and rheumatologic diseases.

Tendons: ultrasound appearance

The sonographic appearance of normal tendons reflects the linear parallel and densely packed arrangement of extracellular collagen [6]. Tendons display a characteristically linear fibrillar arrangement on US when viewed in long axis (Fig. 1). Tendons are highly reflective, because of the strong backscatter of the insonating US beam; consequently, they appear as highly oriented echogenic structures. Because of the arrangement of extracellular collagen, tendon echogenicity is dependent on the angle of insonation of the US beam (Fig. 2). It had been shown that rocking the transducer by as little as 5° to 10° can make the tendon appear hypoechoic, a property referred to as "anisotropy" [7,8]. It is this feature that adds to

---

* Corresponding author.
E-mail address: adlerr@hss.edu (R.S. Adler).

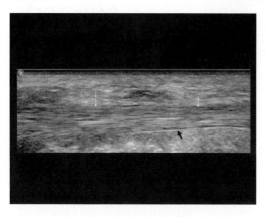

Fig. 1. Longitudinal image of the normal posterior tibial tendon. Normal tendons (*white arrows*) display a characteristically fibrillar architecture when imaged in long axis because of the highly oriented nature of extracellular collagen. The margin of the associated tendon sheath is best seen along the deep surface of the tendon (*black arrow*). A thin hypoechoic line separates it from the overlying tendon.

the difficulty in performing rotator cuff sonography, in which the tendons are curvilinear.

Tendons may be surrounded by either a synovial-lined sheath or dense connective tissue (paratenon). Both structures appear as echogenic borders surrounding the tendon (Figs. 3 and 4). These may be distinguished from the tendon proper by a thin hypoechoic boundary in the case of a tendon sheath. The paratenon does not display anisotropy, which may be helpful in distinguishing it from the adjacent tendon.

On US, tendinosis generally manifests as alterations of tendon morphology and echogenicity [8–10]. Intrasubstance tearing and mucoid degeneration diminish tendon echogenicity (see Fig. 4). Secondary hypertrophy may increase the overall dimension of the tendon. Discrete intrasubstance cystic areas may be apparent. Areas of dystrophic calcification or ossification may be seen, with even tiny foci of calcification being readily appreciated on sonography. Focal calcific masses resulting from calcium hydroxyapatite deposition are likewise well seen on US (Figs. 5 and 6), and may be treated under US guidance [11,12]. The latter generally appear as globular echogenic collections with variable amounts of posterior acoustic shadowing. Tears generally appear as discretely marginated defects within the tendon substance [8]. The presence of surrounding fluid can increase the conspicuity of a tear, whereas the presence of scar, granulation tissue, or calcification may diminish conspicuity. Attritional changes

can be seen as focal or diffuse thinning of a tendon (Figs. 7–9).

As is true for US in general, scan technique is of paramount importance in performing musculoskeletal US. In as much as abnormalities of tendons often appear hypoechoic, maximizing tendon echogenicity diminishes the likelihood of obtaining false-positive results when assessing tendon integrity. This is most evident when examining curvilinear surfaces, such as the shoulder and ankle (Fig. 10).

A number of technical developments have improved the ability to image tendons using US [13–15]. These include tissue-harmonic imaging, compound imaging, and extended field-of-view (FOV) imaging. Tissue-harmonic imaging is a form of nonlinear US imaging that results in improved axial resolution and diminished near-field artifact. Images generally appear cleaner with improved boundary detection. Subtle abnormalities in tendon morphology may become more apparent using tissue-harmonic imaging, as do the margins of tendon tear. Spatial compounding results in speckle reduction, which refers to the inherent graininess seen in US images. Overall tendon morphology and contrast improves with improvement in diagnostic accuracy in assessing tendon abnormalities. Extended-FOV imaging effectively extends the range over which information may be displayed (see Figs. 4 and 9). The full length of a tendon may be imaged from myotendinous junction to the enthesis.

Tendons: MR imaging

The dense fascicles of collagen, which comprise a tendon, are composed of smaller units called microfibrils. The protein matrix of the microfibrils tightly binds water molecules. The microfibrils within a tendon are interwoven in a regular, highly structured fashion [6]. This organization of microfibrils is responsible for the characteristic US appearance of tendons. Similarly, the structured arrangement and tightly bound water molecules causes the characteristic appearance of tendons on MR imaging [16,17]. The lack of mobile water protons causes normal tendons to have low signal intensity on all pulse sequences (Fig. 11).

Tendon imaging on MR imaging

The primary imaging plane used to evaluate tendon pathology is normally the axial plane or trans-

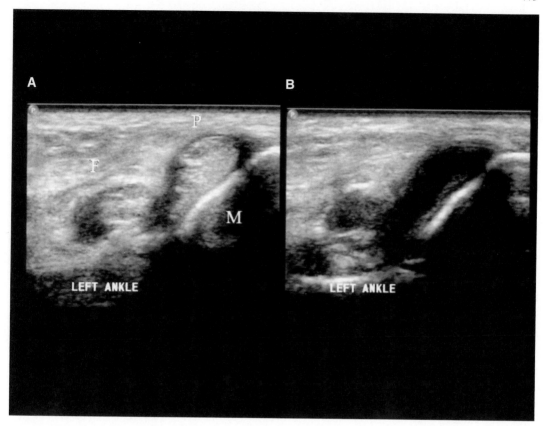

Fig. 2. Anisotropy. (*A*) Transverse image of the normal posterior tibial tendon (P) and flexor digitorum longus tendon (F) obtained at the level of the medial malleolus (M). The tendon is imaged to maximize echogenicity by insonating perpendicular to the tendon axis. (*B*) By rocking the transducer forward or backward relative to the tendon axis, it is noted that both tendons become hypoechoic because of their inherent anisotropy. Although this feature can improve tendon conspicuity, it may also be misinterpreted as tendon pathology.

verse to the long axis of the tendon. One notable exception is the rotator cuff. For most radiologists, the oblique coronal plane is relied on most heavily to identify pathology of the infraspinatus and supraspinatus tendons. For some tendons with a straight course, such as the quadriceps or hamstrings, axial images are perpendicular to the tendon for its entire length (Fig. 12). Although other tendons, such as the peroneal tendons, have a curved course from their myotendinous junctions to their bony insertions, images in a single plane cannot be perpendicular to the tendon for its entire length. Suspected pathology often is confirmed by imaging in an orthogonal plane.

The same principles that guide MR imaging of the musculoskeletal system in general apply to ten-don imaging. Surface coils are routinely used. Thin-section, high-resolution images are necessary to assess tendon pathology accurately. The authors rely heavily on proton density spin echo sequences with long repetition time and short echo time [17].

The presence of free water in otherwise deficient tissues becomes the sine qua non of tissue pathology on MR imaging. This principle applies to tendon pathology but alterations in tendon morphology also must be carefully assessed. Conspicuity of tendon pathology may be enhanced with the use of pre-saturation pulses to take advantage of magnetization transfer effect. In pathologic tendon states, the mobile water content of the tendon increases (Fig. 13). Magnetization transfer effect causes the signal in the normal healthy tendon, which retains its organized

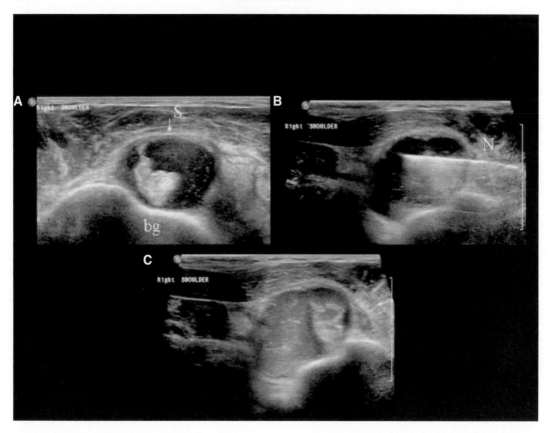

Fig. 3. Tendon sheath injection. (*A*) Transverse image obtained at the level of the bicipital groove (bg) in a patient who has anterior shoulder pain. Hypoechoic fluid and debris surround the biceps tendon. The echogenic tendon sheath (S) is easily identified by virtue of the fluid distention. The biceps tendon is noted to be heterogeneous. (*B*) A 25-gauge needle (N) has been placed into the biceps tendon sheath under US guidance. Because of the advantage of real-time observation, a direct tendon puncture could be avoided. The fluid standoff resulting from the tendon sheath effusion provides a significant advantage in placing the needle. Note the characteristic ring-down artifact displayed by the needle. (*C*) Following injection with a steroid-anesthetic mixture and needle removal, low-level echoes are seen distributed about the biceps tendon. Real-time observation enables one to follow the distribution of the therapeutic mixture during the injection.

structure with little mobile water content, to decrease, causing increased conspicuity of sites with greater mobile water [18]. This principle is not commonly exploited in routine clinical practice, but may be a contributing factor in the use of fast spin echo pulse sequences. Further research in this area may lead to increased use.

Intravenous or intra-articular contrast is not usually required for the evaluation of tendon pathology. It is true, however, that the use of contrast, such as in MR arthrography of the shoulder, may make tendon pathology and specifically rotator cuff tears more conspicuous [19,20]. In general, optimizing scan pa-

rameters to improve contrast and resolution is preferable to the use of contrast material. In the setting where shoulder MR imaging has been performed and a partial cuff tear is identified with a large amount of fluid in the subacromial-subdeltoid bursa, the presence of a complete tear may be strongly suspected. MR arthrography may be helpful to delineate better the full extent of the tear (Fig. 14). The range of tendon pathologies that may be assessed with MR imaging includes myxoid degeneration or tendinosis; partial or complete tears; tenosynovitis; xanthoma formation; depositional arthropathies, such as calcium pyrophosphate or gout with tendon involve-

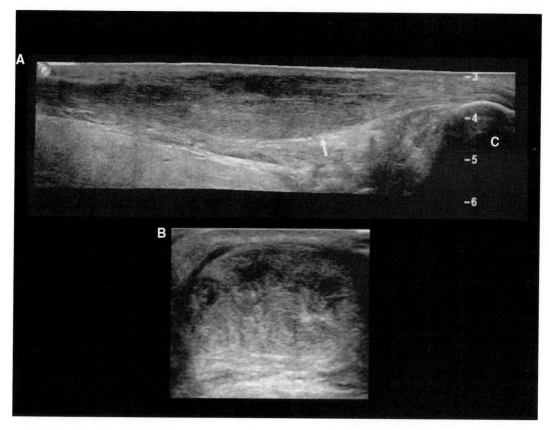

Fig. 4. Achilles tendinosis. (*A*) Extended FOV longitudinal image of a patient who has achillodynia shows fusiform enlargement of most of the central tendon. The tendon is inhomogeneous throughout most of its course. A thin echogenic boundary, best appreciated along the deep surface of the tendon, corresponds to the paratenon (*arrow*), which is closely adherent to the tendon margin. The calcaneus (C) is labeled. (*B*) Better depiction of altered tendon morphology can be appreciated when viewing the tendon in short axis. Multiple intrasubstance clefts and more discrete hypoechoic areas are present. A thin echogenic boundary surrounding the tendon is present corresponding to the paratenon.

ment; or neoplastic states, such as giant cell tumor of the tendon sheath.

Tendinosis is manifest as an ill-defined area of intermediate signal within the tendon reflecting increased free water content (Fig. 15) [21]. In general, in cross-section tendons usually have a round or ovoid appearance. The surface contour is smooth and the tendon caliber is normally uniform throughout its course. Focal thinning or enlargement of the tendon is abnormal. In areas of severe tendinosis, increased signal within the tendon is often accompanied by focal enlargement (Fig. 16). If the surface of the tendon is frayed rather than smooth, this indicates a partial-thickness tear. Complete tendon tears are identified when fluid signal completely disrupts the tendon and the free tendon margins are identified (Fig. 17) [22]. There is some overlap between the

increased signal within a tendon representing tendinosis and the increased signal representing a partial tear. Pathologically, the conditions often coexist (Figs. 18 and 19).

Unlike US, the identification of calcium deposits within a tendon on MR imaging can be very difficult. Calcium hydroxyapatite deposits create a focus of signal void. Unless this causes contour deformity, it can be difficult to recognize when superimposed on the low-signal tendon (Fig. 20). If the calcification has been extruded into an overlying bursa, such as the subacromial-subdeltoid bursa in the shoulder, it may be easier to recognize. A gouty tophus involving a tendon sheath may be clinically manifest as a palpable mass but crystal deposition within a tendon may cause only mild increase in signal, mimicking a partial tendon tear (Fig. 21).

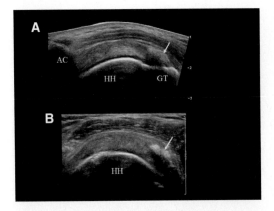

Fig. 5. Calcific tendonitis of the supraspinatus tendon. (A) Long-axis view of the supraspinatus tendon shows an echogenic nodular area with posterior acoustic shadowing at the level of the enthesis (*arrow*). The appearance is characteristic for calcific tendonitis. The superficial margin of the calcific deposit abuts the bursal surface of the tendon. The humeral head (HH) and acromion (AC) are labeled. A thin white line overlying the tendon corresponds to the peribursal fat stripe. (B) Short-axis view of the same tendon shows the calcific deposit (*arrow*) to be situated along the medial aspect of the supraspinatus tendon, adjacent to the rotator interval. GT, greater tuberosity.

When tendons are located in a small space with an adjacent mobile structure, such as within a fibro-osseous tunnel, fascial sling, or beneath a ligamentous band, they usually are enclosed within a covering sheath. This sheath functions to decrease friction between the tendon and the adjacent mobile structure. A small amount of fluid may be physiologic and is readily identified on T2-weighted sequences. Fluid within the sheath is intermediate in signal on proton density sequences and is visible with careful image analysis but it is less conspicuous than on T2-weighted sequences (Fig. 22). Fluid completely surrounding the tendon is usually abnormal indicating tenosynovitis. This may relate to overuse, inflammatory arthritis, or infection.

Care must be taken to avoid misinterpreting all increased signal within a tendon as pathologic. There are two major causes for increased signal within a normal tendon. If multiple tendon slips are converging to form a single tendon, such as the quadriceps, longitudinal striations may be present. The normal tendon has longitudinal bands of intermediate signal within the low-signal tendon [23].

Tendon anisotropy can falsely result in increased signal intensity caused by the so-called "magic an-gle phenomenon" [24]. Diminished rate of T2 relaxation occurs because of a loss of normal dipole–dipole interaction when the tendon is oriented at 55° relative to B_0, the static magnetic field of the scanner. This phenomenon may be seen in tendons with a curved course, such as the supraspinatus tendon or the long flexor or peroneal tendons in the ankle. Because this phenomenon occurs with short echo time sequences, evaluating the tendon on another sequence with a longer echo time or in an orthogonal plane prevents misinterpretation [25].

Complementary roles

It is important to recognize when one modality is better suited than another to address a specific clinical issue, or whether an additional imaging study enhances the information and therapeutic options for a given patient. Common scenarios in which these occur in the authors' institution are the performance of an US-guided intervention following MR imaging, performance of US in lieu of MR imaging because of an indwelling pacemaker, or to observe the dynamic properties of a tendon in the setting of suspected subluxation or impingement. Likewise, the performance of MR imaging is of value when extensive abnormality is suspected, but incompletely characterized by the initial sonographic examination, such as looking for occult fractures or marrow contusions following trauma. The presence of sonographically inaccessible areas, such as those obscured by bone, are examples. It is helpful to emphasize the individual strengths of each modality to determine best which modality is more appropriate, or if a combination of modalities is optimal.

Specific strengths of each modality

Ultrasound

Although protocols exist to examine various anatomic regions (eg, shoulder, ankle, and so forth), US is particularly well-suited to performing targeted examinations, where the area of clinical concern is quite localized. This may, for instance, relate to a specific group of tendons or localized area of tenderness or soft tissue swelling indicated by the patient. Often a diagnosis can be rendered within minutes of performing the examination. In a similar manner, while examining one area, the patient may indicate another

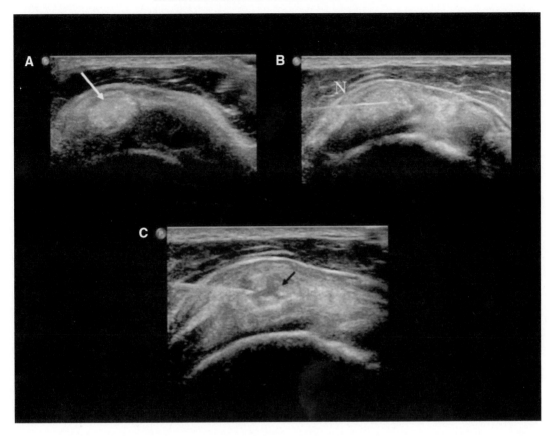

Fig. 6. US-guided therapy of calcific tendonitis. (*A*) Short-axis view of the supraspinatus and portions of the infraspinatus tendon. The transducer has been rocked to make the tendon appear hypoechoic to accentuate the nodular echogenic calcific mass (*arrow*) that projects superiorly into the subdeltoid bursa. The degree of shadowing from these collections can be quite variable and depends primarily on the extent of consolidation of the calcification. In this case a weak posterior acoustic shadow is evident. (*B*) A 20-gauge needle (N) has been placed into the calcification under US guidance. (*C*) During therapeutic lavage and injection of the calcification, a central cavity (*arrow*) within the calcification has become apparent. The calcification is fragmented and lavaged during real-time observation, before injection with an anesthetic-steroid mixture.

area of pain outside the anatomic region requested clinically. It is usually a simple matter to scan the other area of concern indicated by the patient, which in some cases may demonstrate significant related or independent pathology (Fig. 23).

The single most useful and unique aspect of US relates to its real-time nature. This property enables the examiner to perform a series of provocative maneuvers to elicit better suspected pathology. Tendon subluxation and impingement can be demonstrated in this manner (Fig. 24). The combination of real-time examination and the characteristic appearance of metal allow the examiner to visualize a needle for the performance of US-guided interventions [8]. In the authors' institution, US has become

the method of choice to aspirate and inject peritendinous ganglion cysts, tendon sheaths, and areas of calcific tendonitis (see Figs. 3 and 6). US is exquisitely sensitive in detecting even small calcium deposits. It provides a reliable method to perform tendon sheath or peritendinous injections, while avoiding an intratendinous deposition of corticosteroid. The latter has been associated with tendon weakening and possible rupture [26]. Alternatively, situations arise in which more significant pathology becomes apparent only after the introduction of fluid into a bursa or tendon sheath, giving rise to an arthrosonographic effect. Better definition of tendon tears that were occult on initial imaging can impact subsequent therapy (Fig. 25).

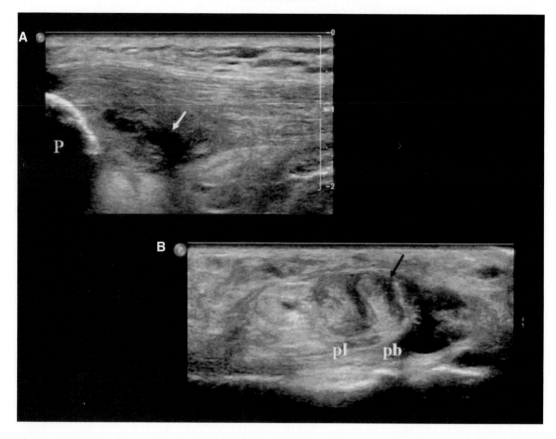

Fig. 7. Partial thickness tears. (*A*) Longitudinal image obtained of the proximal patellar tendon in a patient who has clinically suspected jumper's knee. The proximal patellar tendon is enlarged and inhomogeneous consistent with tendinosis. A discretely marginated hypoechoic area arising within the deep surface of the tendon is present (*arrow*), extending partially into the tendon substance. The patella (P) is labeled. (*B*) Short-axis image of the peroneal tendons below the lateral malleolus in a patient who has lateral ankle pain. The peroneus brevis (pb) and longus (pl) are seen in cross-section in this view. Both tendons are heterogeneous with the former displaying a prominent longitudinal split tear arising from its superficial margin (*arrow*).

Vascular assessment is possible using Doppler techniques without the necessity of performing an intravenous injection of contrast. In as much as symptomatic tendinosis can manifest an angiofibroblastic response [27,28], the presence of abnormal intratendinous and peritendinous hyperemia is of clinical significance (Fig. 26). The presence of hyperemia may help localize areas of the tendon that may respond to peritendinous or intratendinous anesthetic or steroid injections. Likewise, active peritendinous inflammation occurs in the setting of an adjacent tenosynovitis or bursitis [29], as may occur in rheumatoid arthritis (Fig. 27). Revascularization is also an important factor in healing [30]. It has been shown that early following rotator cuff repair there is marked

hyperemic response at or near the repair site, which diminishes over time in a predictable manner (Fig. 28). US can provide an accessible and cost-effective method to assess response to therapeutic interventions, which can manifest as diminished vascularity over time.

The presence of metallic hardware or other implants, in particular metal plates and screws, does not preclude sonographic evaluation of tendons. Indeed, metal has a characteristic sonographic signature, appearing as a strong specular surface with a prominent posterior ring-down artifact [31]. A frequently requested examination in the authors' institution is to assess the integrity of a tendon that courses near a metallic screw, pin, or plate, both as static images and

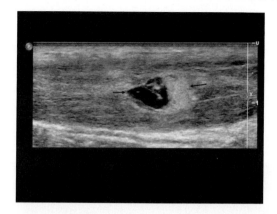

Fig. 8. Intrasubstance tear of the Achilles tendon. Longitudinal image of the mid substance of the tendon shows it to be enlarged and inhomogeneous with ill-defined margins. A central discretely marginated abnormality within the tendon corresponds to a tear identified on MR imaging. The complex nature of the tear (*arrows*) shows it to be filled with both fluid and soft tissue, presumably granulation tissue. US-guided decompression of the fluid-filled cavity was performed resulting in nonviscous clear material. A peritendinous therapeutic injection was performed for temporary pain relief.

with US examination, may be altered if associated subtalar joint arthrosis and injury of the spring ligament is recognized on MR imaging examination [32].

In some regions, bony anatomy may restrict access for evaluation by a US transducer (Fig. 34). Tears of the myotendinous junction of the rotator cuff are rare and quite difficult to diagnose with US but are readily amenable to MR imaging diagnosis.

Even though MR imaging is extremely useful for imaging an entire region, the region that can be evaluated is restricted to the area covered by the surface coil. Occasionally, this may be problematic. For instance, if a patient has a quadriceps rupture in the knee and knee MR imaging is performed using a standardized protocol, the entire tendon may not be included. Management of the injury requires knowledge of the amount of retraction of the tendon and some assessment of the tendon margin. This may require identification of the pathology by the technologist performing the examination and adjustment of the coil position to include the retracted tendon (Figs. 35 and 36). Although MR imaging is considerably less operator dependent in general as compared with US, there are some instances where adjustments to routine are required.

dynamically (Fig. 29). Patients who have indwelling metallic implants and pacemakers that may preclude the possibility of performing MR imaging can often be imaged using US as a viable alternative (Fig. 30). Likewise, patients who are claustrophobic and are not candidates for MR imaging can benefit from US examination. Combinations of these features generally apply in performing US. One does not preclude the other. The presence of a screw from an osteotomy or fracture reduction may impinge an adjacent tendon and alter its normal function, requiring dynamic assessment. An adjacent inflammatory response may benefit from US-guided therapy.

MR imaging

The greatest strength of MR imaging lies in its ability globally to assess the anatomy of a region. The ability to detect not only tendon pathology but also bony abnormalities, cartilage injuries or sites of cartilage loss, and ligament damage is extremely useful in guiding patient management (Figs. 31–33). For example, management of the patient who has a partial posterior tibial tendon tear, which may be diagnosed

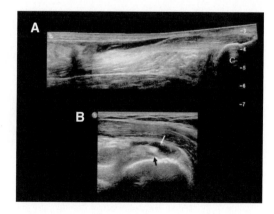

Fig. 9. Full-thickness tears. (*A*) Longitudinal extended FOV image of the Achilles tendon shows a large (complete) full-thickness tear originating at the myotendinous junction and extending approximately 3 cm. The relationship of the tear to the tendon origin and insertion and degree of retraction are well-depicted in this extended view. Hypoechoic fluid fills an intact paratenon. The calcaneus (C) is labeled. (*B*) Long-axis image of the supraspinatus tendon shows distention of the subdeltoid bursa (*white arrow*) by fluid, which also fills a gap at the tendon footprint. Fluid slightly undercuts the free end of the tendon, extending partially over the articular cartilage (*black arrow*).

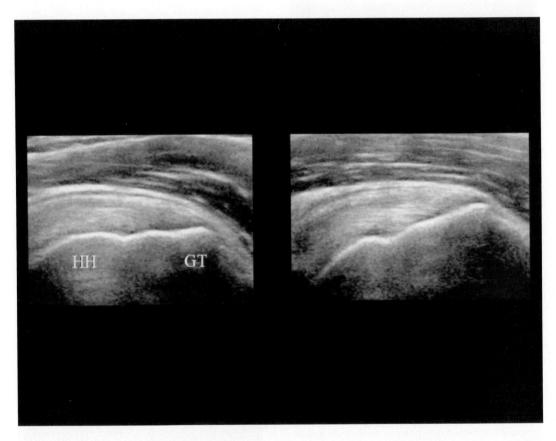

Fig. 10. Normal supraspinatus tendon: effect of anisotropy. The echogenicity of curved tendon surface may vary because of differences in transducer orientation. In the left image, the proximal supraspinatus tendon is imaged to maximize echogenicity, whereas the footprint is suboptimally imaged. In the right-sided image, the transducer has been rocked to improve visualization of the tendon footprint, whereas the more proximal tendon has become less echogenic. The humeral head (HH) and greater tuberosity (GT) are labeled.

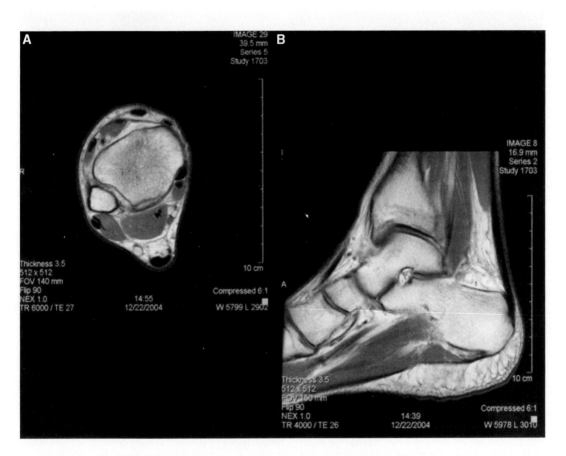

Fig. 11. Normal anatomy and normal appearance of the ankle tendons. (*A*) In cross-section the anterior margin of the Achilles tendon is flat or concave. (*B*) On sagittal images the tendon has low signal and uniform caliber.

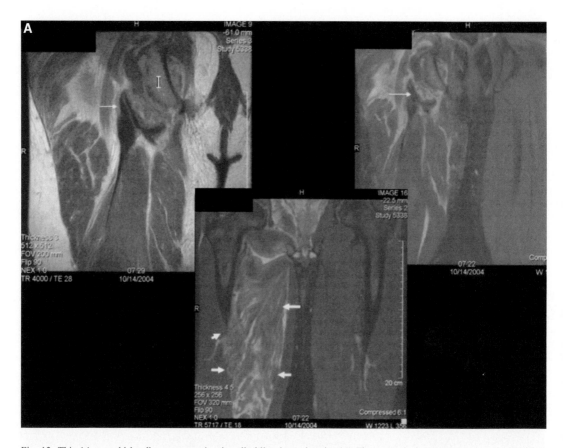

Fig. 12. This 14-year-old hurdler was experiencing disabling buttock pain. (*A*) The coronal plane demonstrates the avulsion of the ischial apophysis with the attached hamstring tendons (*thin white arrow*). The gap between the apophysis and the ischium can be measured and the very extensive edema-hemorrhage in the hamstring musculature and adductor magnus can be appreciated (*thick white arrows*). (*B*) The axial images confirm the integrity of the hamstring tendons. The adjacent sciatic nerve is surrounded by hemorrhage proximally (*arrow*).

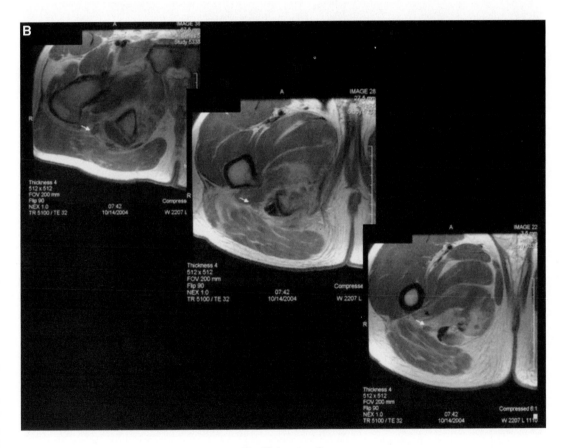

Fig. 12 (*continued*).

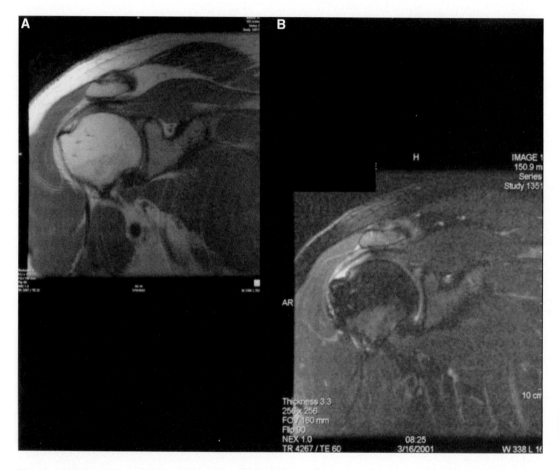

Fig. 13. (A, B) The extent of this partial articular surface tear of the supraspinatus tendon is easier to define on the water-sensitive fat-suppressed proton density (PD) sequence (B).

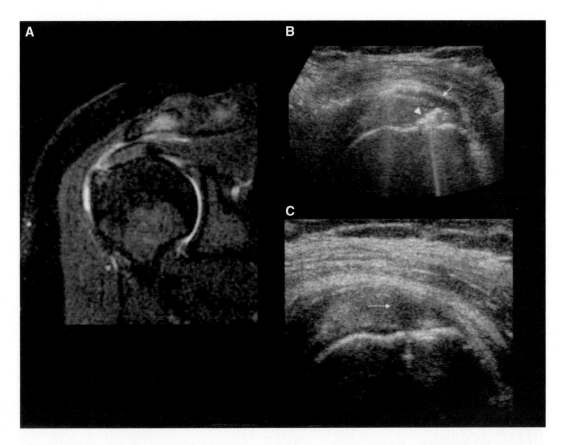

Fig. 14. (*A*) Initially MR image was interpreted as fraying of the supraspinatus tendon with small bursal surface tear. The patient was referred to US for a therapeutic injection. (*B*) Arthrosonographic effect of instilled steroid solution following intra-articular injection makes full-thickness tear more conspicuous. Microbubbles (*arrowhead*) of instilled steroid solution cross full-thickness tear and fluid extends to subdeltoid-subacromial bursa (*arrow*). (*C*) Initial US evaluation suggests a partial articular surface tear (*arrow*).

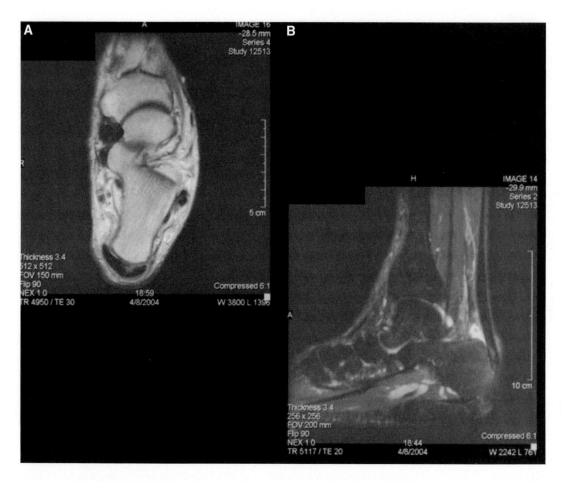

Fig. 15. Achilles tendinosis. Axial PD (*A*) and sagittal fast spin echo inversion recovery (FSE IR) (*B*) images show the Achilles tendon is enlarged with areas of intermediate signal. There is associated deep retrocalcaneal bursitis.

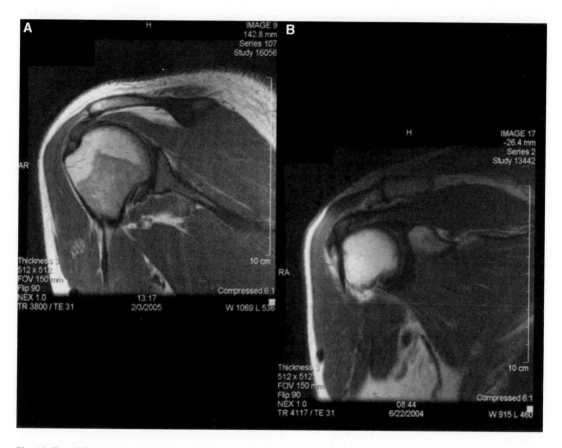

Fig. 16. Two different patients who have shoulder pain. (*A*) The infraspinatus tendon is enlarged with increased signal indicating severe tendinosis. (*B*) Severe tendinosis of the supraspinatus.

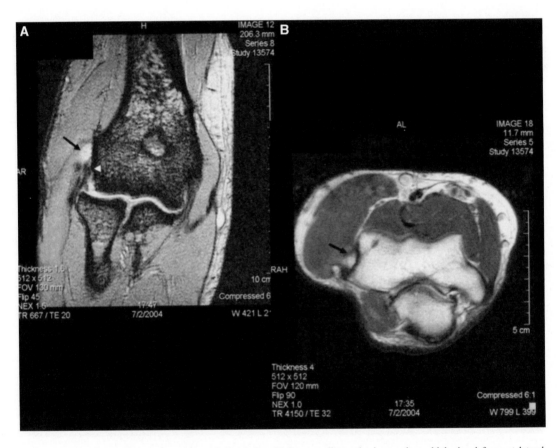

Fig. 17. Patient who has lateral elbow pain. (*A*) Coronal multiplanar gradient echo image shows high signal focus at lateral humeral condyle caused by avulsion of the extensor carpi radialis brevis (*arrow*). The lateral collateral ligament remains intact (*arrowhead*). (*B*) Axial PD supports diagnosis of extensor carpi radialis brevis avulsion (*arrow*).

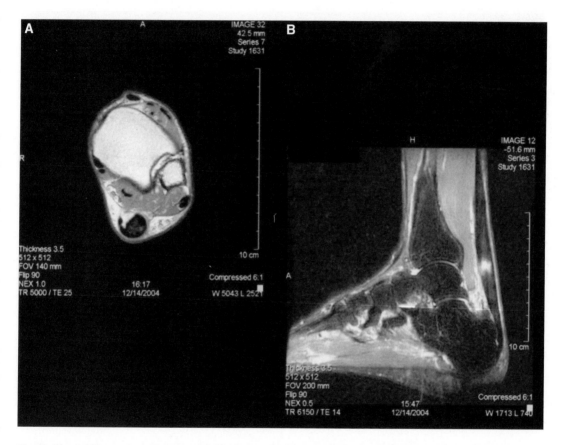

Fig. 18. The Achilles tendon shows a long segment of tendinosis and a more focal partial tear. Axial PD image (*A*) shows tendon enlargement with intermediate signal corresponding to tendinosis. Tendinosis is more conspicuous on sagittal FSE IR (*B*). The stellate focus of very bright signal within the long segment is the site of partial tear.

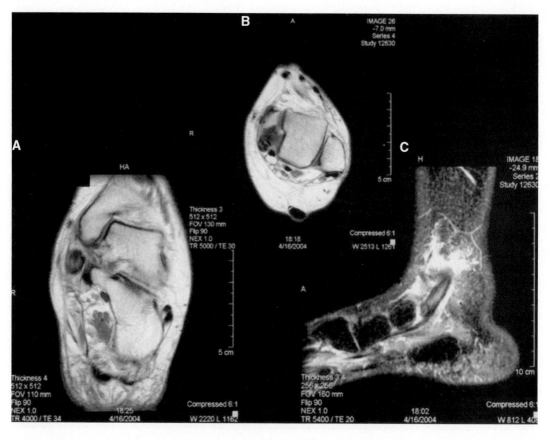

Fig. 19. Patient who has medial ankle pain. There is enlargement of the posterior tibial tendon at the level of the medial malleolus and extending distally. (*A*) Coronal PD image shows enlarged posterior tibial tendon with increased signal. (*B*) Axial PD image shows rounded foci of brighter signal indicating intrasubstance split. (*C*) Sagittal FSE IR image shows severe tendinosis, intrasubstance split, and extensive peritendinous edema.

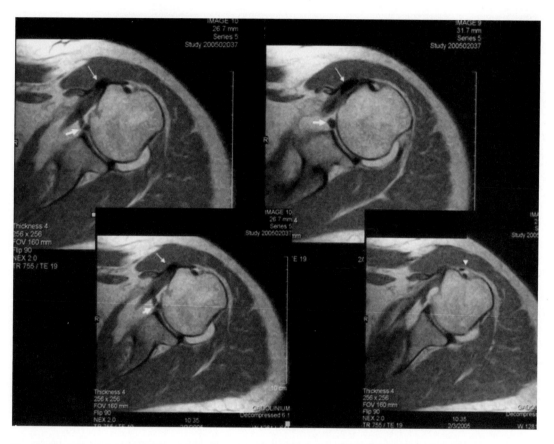

Fig. 20. This patient who had shoulder pain underwent MR arthrography. Calcium deposition in the subscapularis tendon can be identified because the ovoid focus distorts the tendon contour and is lower in signal than the degenerated tendon (*thin arrow*). A displaced anterior labral tear is also present (*thick arrow*). Anatomic variants, such as the accessory biceps, may be easier to identify with MR arthrography (*arrowhead*).

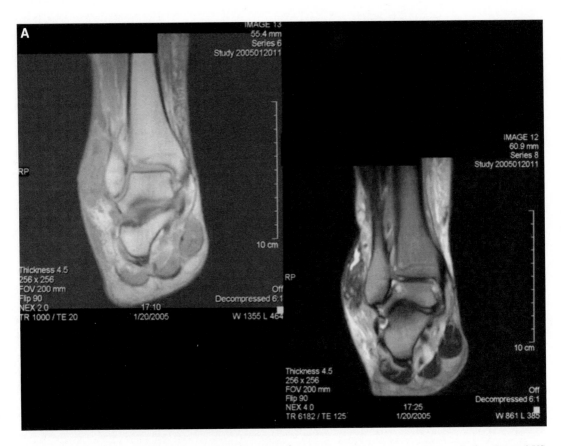

Fig. 21. (*A, B*) This 52-year-old man presented with a painful palpable mass adjacent to the lateral malleolus. Targeted US examination may have made the diagnosis of gout because of its characteristic appearance. MR imaging examination, however, showed the extent of disease. Bulky synovitis involving the ankle joint, the talonavicular joint, the first metatarsophalangeal joint, osseous erosions, and tophaceous deposits involving multiple tendon sheaths including the tibialis anterior and the extensor digitorum are shown.

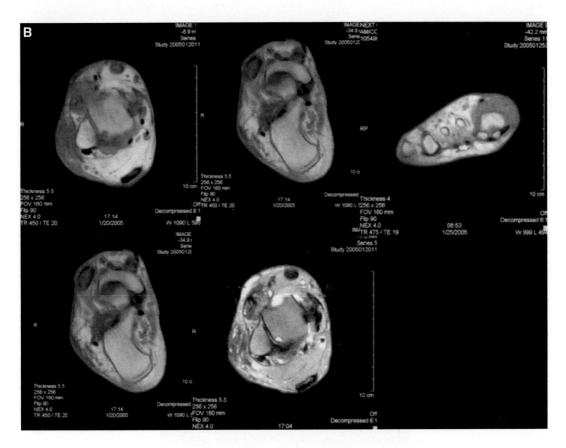

Fig. 21 (*continued*).

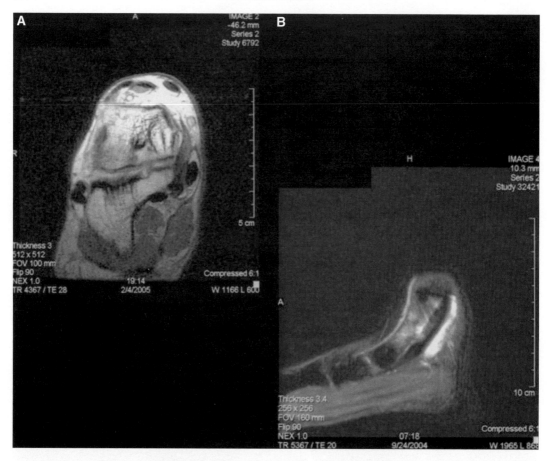

Fig. 22. (*A*) Posterior tibial tenosynovitis on coronal PD image. (*B*) Posterior tibial tenosynovitis on IR sequence is more conspicuous.

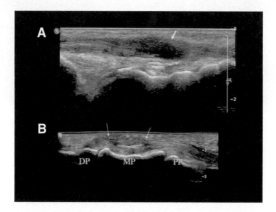

Fig. 23. (*A*) Patient complaining of pain and swelling over the dorsal and medial aspect of forefoot. Scans over the area of localized pains demonstrate a complete rupture of the tibialis anterior tendon. The palpable abnormality corresponds to the region of tendon retraction, which appears bulbous and hypoechoic (*arrow*). (*B*) Patient who has history of recent flexor tendon laceration and repair is unable to flex at the distal interphalangeal joint following aggressive therapy. Extended FOV image shows the volar surface of the digit, including proximal and distal interphalangeal joints. A complete dehiscence of the repair is evident at the level of the middle phalanx (*arrows*). The proximal stump is retracted to the base of the phalanx and the distal stump is attached to bone. DP, distal phalanx; MP, middle phalanx; PP, proximal phalanx.

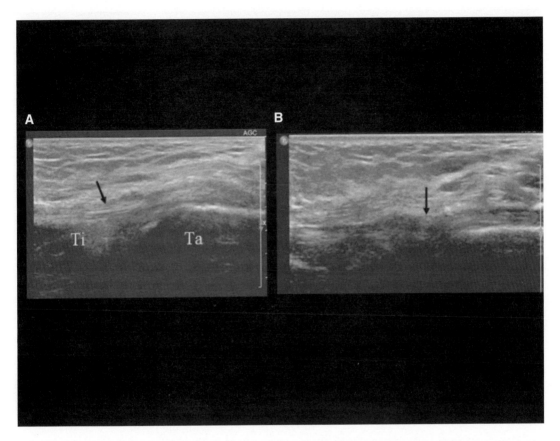

Fig. 24. Ballet dancer who experiences pain with plantar flexion and "pointing" over the posteromedial aspect of the ankle. (*A*) Long-axis image of the flexor hallucus longus tendon (*arrow*) imaged at the level of the posterior tibiotalar joint with the foot in a neutral position. The posterior margins of the tibia (Ti) and talus (Ta) are labeled. (*B*) With plantar flexion, the tendon should normally glide smoothly within its sulcus. In this case, a prominent kink (*arrow*) became evident with restricted gliding motion of the tendon at the level of the posterior tibiotalar joint, presumably related to scarring or adhesions.

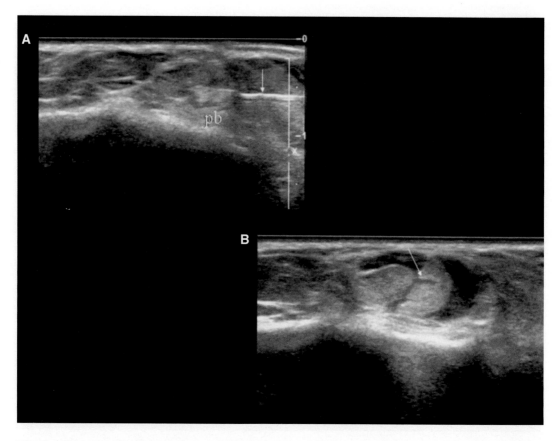

Fig. 25. Arthrosonographic effect depicting subtle tear in a patient referred for US-guided tendon sheath injection following diagnosis of tendinosis by MR imaging. (*A*) Short-axis image of inframaleolar peroneal tendons following placement of a 25-gauge needle. The echogenic needle (*arrow*) is seen abutting the margin of the peroneus brevis. Both peroneal tendons appear heterogeneous. (*B*) Following injection and needle removal, the tendon sheath is distended improving the definition of the tendon margins. Improved visualization of the tendon margins enables diagnosis of a partial-thickness split (*arrow*) arising from the posterior margin of the peroneus brevis that was not evident on MR imaging or initial gray-scale US.

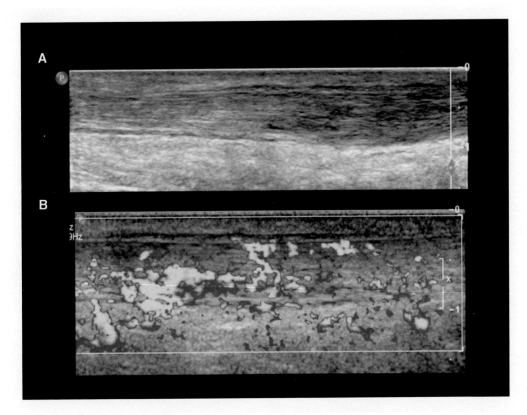

Fig. 26. Severe achillodynia. (*A*) Long-axis image of the Achilles tendon in a patient who has severe retrocalcaneal pain. The tendon is enlarged and inhomogeneous. (*B*) There is marked peritendinous and intratendinous blood flow on power Doppler imaging reflecting the extensive angiofibroblastic response.

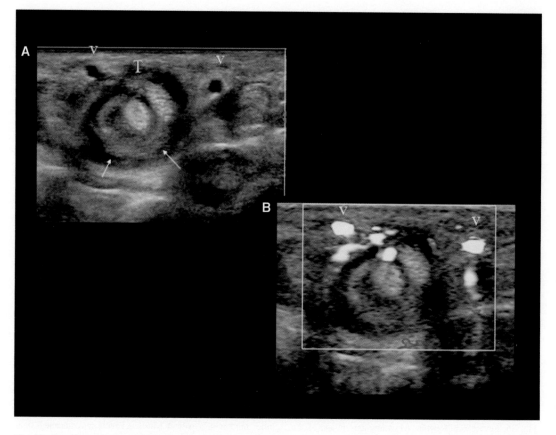

Fig. 27. Tenosynovitis in rheumatoid arthritis. (*A*) Short-axis view of the flexor tendons (T) of the third digit in the palmar aspect of the hand shows abnormal peritendinous soft tissue (*arrows*). Two interdigital vessels (v) are evident. (*B*) Hyperemia within and around the sheath on power Doppler imaging is indicative of active disease.

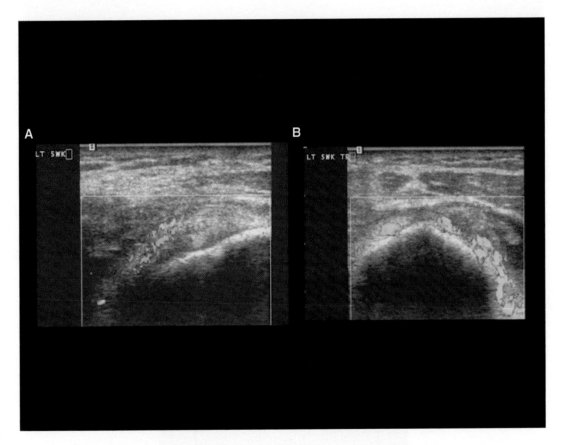

Fig. 28. Long- (*A*) and short- (*B*) axis views of the supraspinatus tendon obtained 5 weeks following rotator cuff repair. There is prominent vascularity depicted both within the tendon (*A*) and in a pericortical (*B*) distribution. This has been shown to be a fairly consistent finding in the early postoperative period, following cuff repair.

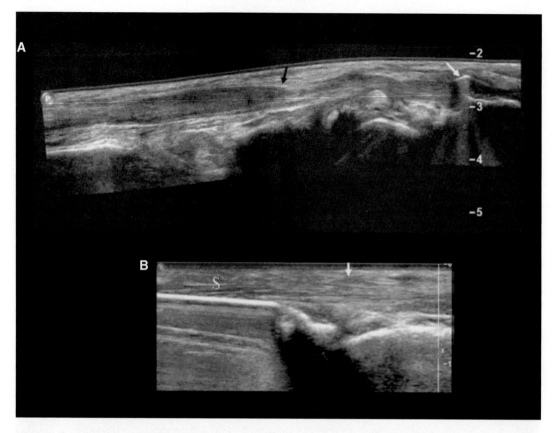

Fig. 29. Peroneus longus rupture caused by metallic staple. (*A*) Extended FOV longitudinal image of the peroneal tendons shows a complete rupture with retraction of the peroneus longus tendon (*black arrow*). The peroneus brevis is in continuity and in close relationship to a metallic staple projecting from the lateral margin of the calcaneus (*white arrow*). (*B*) A longitudinal view of the staple shows its relation to the peroneus brevis tendon (*arrow*). Note the reverberation artifact from the staple (S), which does not interfere with assessment of the adjacent structures.

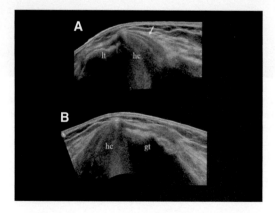

Fig. 30. Total shoulder replacement with high riding prosthesis on radiographs. (*A*) Extended FOV imaging over the anterior shoulder obtained with the patient in external rotation. The metallic humeral component (hc) and adjacent bone (lt) are readily distinguished as are the overlying soft tissue planes. In this case no discernable subscapularis tendon is evident, with the metallic prosthesis impinging the subdeltoid fat (*arrow*). There is complete atrophy of deltoid muscle. (*B*) Extended FOV long-axis image over the greater tuberosity (gt) again shows absence of residual cuff tissue and superior migration of the humeral component (hc).

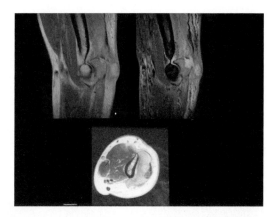

Fig. 31. Patient who has elbow swelling underwent US examination, which demonstrated olecranon bursitis and surface erosion of the olecranon. MR imaging was suggested for further evaluation to assess extent of bone involvement. MR imaging (sagittal PD, sagittal IR, axial PD) showed myositis of the triceps with partial rupture of the triceps tendon. Osteomyelitis was demonstrated with the infected olecranon bursitis eroding the olecranon. Extensive reactive marrow edema surrounds the olecranon erosion.

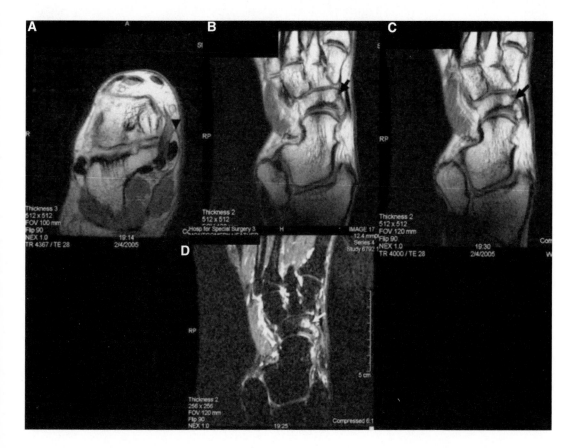

Fig. 32. This 29-year-old runner had been having medial mid-foot pain for a few months. Although US examination would have defined the posterior tibial tenosynovitis (*arrowhead*), MR imaging defined more extensive pathology. Oblique, coronal PD, and long-axis PD images (*A–C*) show an ununited fracture of the medial navicular. The long-axis FSE IR image (*D*) shows cystic changes at the interface (*arrow*). Marked thickening of the subchondral plate of the proximal navicular surface is present with increased signal in the overlying cartilage and surface fibrillation. There is also hyperintensity of the cartilage of the medial aspect of the anterior process of the talus with focal depression of the subchondral plate (*white arrow*).

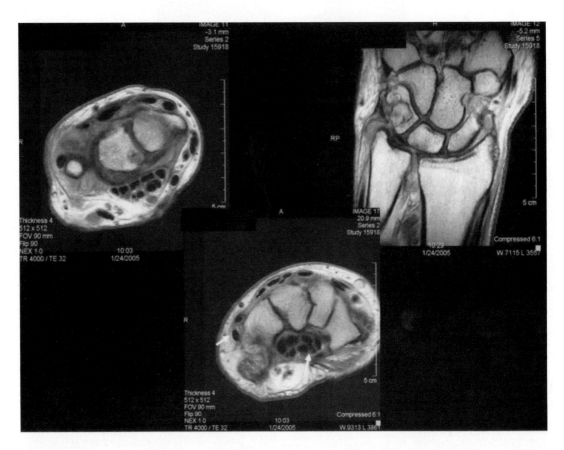

Fig. 33. This 56-year-old man presented to his physician with a swollen wrist. MR imaging examination (axial and coronal PD) shows diffuse synovitis with disruption of the triangular fibrocartilage complex. There are osseous erosions of the scaphoid and triquetrum. Partial splits of the extensor carpi ulnaris and flexor indicis superficialis are demonstrated (*arrows*).

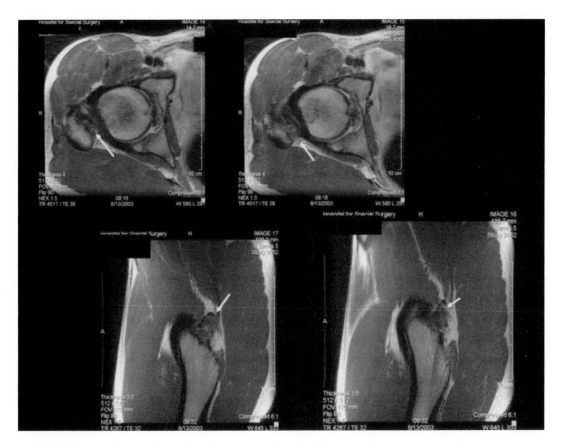

Fig. 34. This 17-year-old sustained an avulsion of the insertion of the pyriformis from the posterior greater trochanter with subsequent tendinosis and dystrophic ossification (*arrow*). Bony anatomy makes this area difficult to evaluate with US for primary diagnosis. Subsequently, patient had therapeutic injection performed with US guidance and did well.

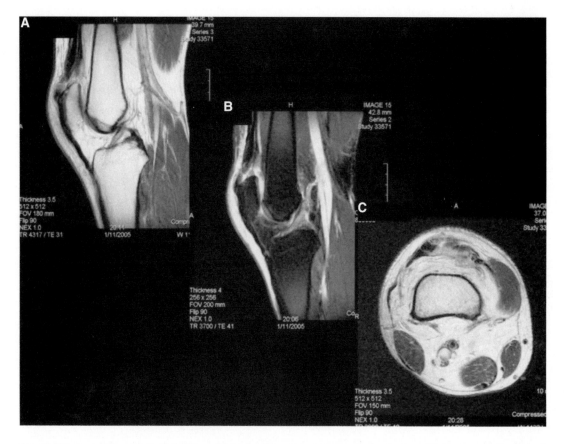

Fig. 35. This young man presented after "bumping his knee on a door." He sustained a distal quadriceps rupture with minimal gap between the tendon fragments. (*A*, *B*) Focal tendon disruption is evident on sagittal imaging (PD and IR, respectively). (*C*) Loss on normal tendon morphology is best appreciated on axial imaging.

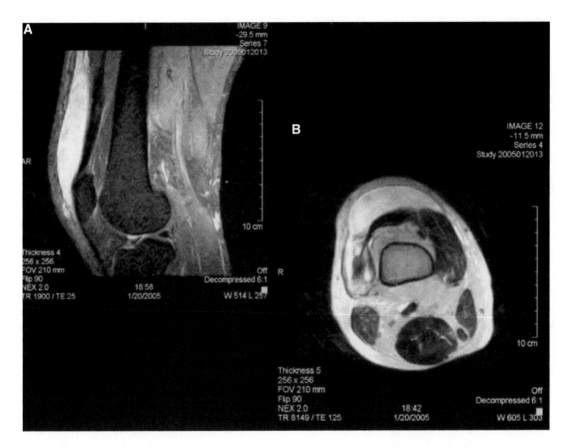

Fig. 36. This patient was unable to straight leg raise after a fall and a quadriceps tear was suspected. The surface coil was positioned more proximally than for a standard knee examination. Sagittal fat-suppressed image (*A*) and axial T2-weighted image (*B*) are presented. The patient's extensor mechanism was intact but he did have a very large hematoma anterior to the quadriceps.

Summary

The choice to use MR imaging or US to depict tendon pathology has traditionally depended on the imager's level of experience and comfort with the modality, and, less so, the individual strengths of either modality. Although this may be an acceptable rationale, it does not fully take advantage of the strength of either modality or the potential benefits of combining both modalities. Musculoskeletal imaging is truly a multimodality approach to defining pathology, so that being proficient in one area of imaging should not preclude the need to become adept at others. This article elaborates on the authors' experience in an institution where both modalities are highly used and have become part of the algorithm to address a large variety of both tendon and general musculoskeletal issues. As such, a significantly broader range of tendon pathology may be addressed within the context of an imaging department.

References

[1] Davies SG, Baudouin J, King JB, et al. Ultrasound, computed tomography and magnetic resonance imaging in patellar tendonitis. Clin Radiol 1991;43:52–6.

[2] Teefey SA, Middleton WD, Rubin DA, et al. Detection of partial and full thickness rotator cuff tears in patients with a painful shoulder: a comparison of ultrasound, MRI and arthroscopic surgery. Radiology 2000;217(Suppl):430.

[3] Bryant L, Shnier R, Bryant C, et al. A comparison of clinical estimation, ultrasonography, magnetic resonance imaging and arthroscopy in determining the size of rotator cuff tear. J Shoulder Elbow Surg 2002;11: 219–24.

[4] Chang CY, Wang SF, Chiou HJ, et al. Comparison of shoulder ultrasound and MR imaging in diagnosing full thickness rotator cuff tears. Clin Imaging 2002; 26:50–4.

[5] Teefey SA, Rubin DA, Middleton WD, et al. Detection and quantification of rotator cuff tears: a comparison of ultrasound, MRI and arthroscopic findings in seventy one consecutive cases. J Bone Joint Surg Am 2004; 86:708–16.

[6] Elliott DH. Structure and function of mammalian tendons. Biol Rev 1965;40:392–421.

[7] Bouffard JA, Eyler WR, Introcaso JH, et al. Sonography of tendons. Ultrasound Quarterly 1993;11: 259–86.

[8] Adler RS, Sofka CM. Percutaneous ultrasound-guided injections in the musculoskeletal system. Ultrasound Quarterly 2003;19:3–12.

[9] Kainberger FM, Engel A, Barton P, et al. Injury to the Achilles tendon: diagnosis with sonography. AJR Am J Roentgenol 1990;155:1031–6.

[10] Sell S, Schulz R, Balentsiefen M, et al. Lesions of the Achilles tendon: a sonographic, biochemical and histological study. Arch Orthop Trauma Surg 1996; 115:28–32.

[11] Farin PU, Jaroma H, Soimakallio S. Rotator cuff calcifications: treatment with ultrasound-guided technique. Radiology 1995;195:841–3.

[12] Howard CB, Vinzberg A, Nyska M, et al. Aspiration of acute calcerous trochanteric bursitis using ultrasound guidance. J Clin Ultrasound 1993;21:45–7.

[13] Barberie JE, Wong AD, Cooperberb PL, et al. Extended field-of-view sonography in musculoskeletal disorders. AJR Am J Roentgenol 1998;171:751–7.

[14] Adler RS. Future and new developments in musculoskeletal ultrasound. Radiol Clin North Am 1999;27: 623–31.

[15] Lin DC, Nazarian LN, O'Kane PL, et al. Advantages of real-time spatial compound sonography of the musculoskeletal system versus conventional sonography. AJR Am J Roentgenol 2002;179:1629–31.

[16] Beltran J, Noto AM, Herman LJ, et al. Tendons: high field strength surface coil imaging. Radiology 1987; 162:735–40.

[17] Koblik PD, Freeman DM. Short echo time magnetic resonance imaging of tendon. Invest Radiol 1993;28: 1095–100.

[18] Adler R, Swanson S, Doi K, et al. The effect of magnetization transfer in meniscal fibrocartilage. Magn Reson Med 1996;35:591–5.

[19] Flannigan B, Kursonoglu-Brajme S, Snyder S, et al. MR arthrography of the shoulder: comparison with conventional MR imaging. AJR Am J Roentgenol 1990; 155:829–32.

[20] Yagci B, Manisali M, Yilmaz E, et al. Indirect MR arthrography of the shoulder in detection of rotator cuff ruptures. Eur Radiol 2001;11:258–62.

[21] Movin T, Kristoffersen-Wiberg M, Rolf C, et al. MR imaging in chronic Achilles tendon disorder. Acta Radiol 1998;39:126.

[22] Schweitzer ME, Karasick D. MR imaging of disorders of the Achilles tendon. AJR Am J Roentgenol 2000; 175:613.

[23] Zeiss J, Saddemi SR, Ebraheim NA. MR imaging of the quadriceps tendon: normal layered configuration and its importance in cases of tendon rupture. AJR Am J Roentgenol 1992;159:1031–4.

[24] Erickson SJ, Cox IH, Hyde JS, et al. Effect of tendon orientation on MR imaging signal intensity: a manifestation of the "magic angle" phenomenon. Radiology 1991;181:389.

[25] Peh WC, Chan JH. The magic angle phenomenon in tendons: effect of varying MR echo time. Br J Radiol 1998;71:31.

[26] Unverferth LJ, Olix ML. The effect of local steroid injection on tendon. J Sports Med 1973;1(4):31–7.

[27] Astrom M, Westlin M. Blood flow in chronic Achilles tendinopathy. Clin Orthop 1994;308:166–72.

[28] Zanetti M, Metzdorf A, Kundert HP, et al. Achilles Tendon: clinical relevance of neovascularization diagnosed with power Doppler US. Radiology 2003;227: 556–60.

[29] Newman JS, Adler RS, Bude RO, et al. Detection of soft tissue hyperemia: value of power Doppler sonography. AJR Am J Roentgenol 1994;163:385–9.

[30] Adler RS, Fealy S. Ultrasound of rotator cuff tear: current status. Techniques in Shoulder & Elbow Surg 2003;4:121–32.

[31] Yasher AA, Adler RS, Grady-Benson JC, et al. Ultrasound method to evaluate polyethylene component wear in total knee replacement arthroplasty. Am J Orthop 1996;25:702–4.

[32] Balen P, Helms C. Association of posterior tibial tendon injury with spring ligament injury, sinus tarsi abnormality and plantar fasciitis on MR imaging. AJR Am J Roentgenol 2001;176:1137–43.

ELSEVIER
SAUNDERS

Radiol Clin N Am 43 (2005) 809–814

**RADIOLOGIC
CLINICS**
of North America

Index

Note: Page numbers of article titles are in **boldface** type.

Changing Your Address?

Make sure your subscription changes too! When you notify us of your new address, you can help make our job easier by including an exact copy of your Clinics label number with your old address (see illustration below.) This number identifies you to our computer system and will speed the processing of your address change. Please be sure this label number accompanies your old address and your corrected address—you can send an old Clinics label with your number on it or just copy it exactly and send it to the address listed below.

We appreciate your help in our attempt to give you continuous coverage. Thank you.

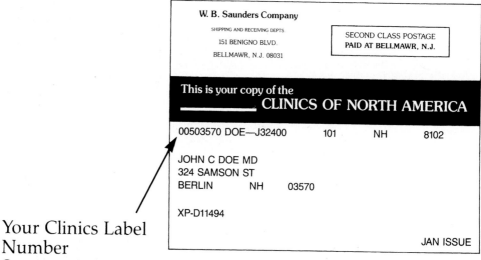

Your Clinics Label Number
Copy it exactly or send your label along with your address to:
W.B. Saunders Company, Customer Service
Orlando, FL 32887-4800
Call Toll Free 1-800-654-2452

Please allow four to six weeks for delivery of new subscriptions and for processing address changes.

Practice, Current, Hardbound:
SATISFACTION GUARANTEED

Adolescent Medicine Clinics
❏ Individual $95
❏ Institutions $133
❏ *In-training $48

Anesthesiology
❏ Individual $175
❏ Institutions $270
❏ *In-training $88

Cardiology
❏ Individual $170
❏ Institutions $266
❏ *In-training $85

Chest Medicine
❏ Individual $185
❏ Institutions $285

Child and Adolescent Psychiatry
❏ Individual $175
❏ Institutions $265
❏ *In-training $88

Critical Care
❏ Individual $165
❏ Institutions $266
❏ *In-training $83

Dental
❏ Individual $150
❏ Institutions $242

Emergency Medicine
❏ Individual $170
❏ Institutions $263
❏ *In-training $85
❏ Send CME info

Facial Plastic Surgery
❏ Individual $199
❏ Institutions $300

Foot and Ankle
Individual $160
Institutions $232

Gastroenterology
❏ Individual $190
❏ Institutions $276

Gastrointestinal Endoscopy
❏ Individual $190
❏ Institutions $276

Hand
❏ Individual $205
❏ Institutions $319

Heart Failure (NEW in 2005!)
❏ Individual $99
❏ Institutions $149
❏ *In-training $49

Hematology/ Oncology
❏ Individual $210
❏ Institutions $315

Immunology & Allergy
❏ Individual $165
❏ Institutions $266

Infectious Disease
❏ Individual $165
❏ Institutions $272

Clinics in Liver Disease
❏ Individual $165
❏ Institutions $234

Medical
❏ Individual $140
❏ Institutions $244
❏ *In-training $70
❏ Send CME info

MRI
❏ Individual $190
❏ Institutions $290
❏ *In-training $95
❏ Send CME info

Neuroimaging
❏ Individual $190
❏ Institutions $290
❏ *In-training $95
❏ Send CME inf0

Neurologic
❏ Individual $175
❏ Institutions $275

Obstetrics & Gynecology
❏ Individual $175
❏ Institutions $288

Occupational and Environmental Medicine
❏ Individual $120
❏ Institutions $166
❏ *In-training $60

Ophthalmology
❏ Individual $190
❏ Institutions $325

Oral & Maxillofacial Surgery
❏ Individual $180
❏ Institutions $280
❏ *In-training $90

Orthopedic
❏ Individual $180
❏ Institutions $295
❏ *In-training $90

Otolaryngologic
❏ Individual $199
❏ Institutions $350

Pediatric
❏ Individual $135
❏ Institutions $246
❏ *In-training $68
❏ Send CME info

Perinatology
❏ Individual $155
❏ Institutions $237
❏ *In-training $78
❏ Send CME inf0

Plastic Surgery
❏ Individual $245
❏ Institutions $370

Podiatric Medicine & Surgery
❏ Individual $170
❏ Institutions $266

Primary Care
❏ Individual $135
❏ Institutions $223

Psychiatric
❏ Individual $170
❏ Institutions $288

Radiologic
❏ Individual $220
❏ Institutions $331
❏ *In-training $110
❏ Send CME info

Sports Medicine
❏ Individual $180
❏ Institutions $277

Surgical
❏ Individual $190
❏ Institutions $299
❏ *In-training $95

Thoracic Surgery (formerly Chest Surgery)
❏ Individual $175
❏ Institutions $255
❏ *In-training $88

Urologic
❏ Individual $195
❏ Institutions $307
❏ *In-training $98
❏ Send CME info

Order your subscription today. Simply complete and detach this card and drop it in the mail to receive the best clinical information in your field.

BUSINESS REPLY MAIL

FIRST-CLASS MAIL   PERMIT NO 7135   ORLANDO FL

POSTAGE WILL BE PAID BY ADDRESSEE

PERIODICALS ORDER FULFILLMENT DEPT
ELSEVIER
6277 SEA HARBOR DR
ORLANDO FL 32821-9816